Ultraviolet Spectra of Elastomers and Rubber Chemicals

Ultraviolet Spectra of Elastomers and Rubber Chemicals

V. S. Fikhtengol'ts, R. V. Zolotareva,
and Yu. A. L'vov

All-Union Synthetic Rubber Research Institute, Leningrad

Translated from Russian by
A. Eric Stubbs
Senior Scientific Translator

SPRINGER SCIENCE+BUSINESS MEDIA, LLC

1966

Library of Congress Catalog Card Number 66-12889

The original Russian text was published for the All-Union Synthetic Rubber Research Institute by Khimiya Press in Moscow in 1965.

ULTRAVIOLET SPECTRA OF ELASTOMERS AND RUBBER CHEMICALS

ATLAS UL'TRAFIOLETOVYKH SPEKTROV POGLOSHCHENIYA VESHCHESTV, PRIMENYAYUSHCHIKHSYA V PROIZVODSTVE SINTETICHESKIKH KAUCHUKOV

АТЛАС ультрафиолетовых спектров поглощения веществ, применяющихся в производстве синтетических каучуков

В. С. Фихтенгольц, Р. В. Золотарева,
Ю. А. Львов

ISBN 978-1-4615-9593-9   ISBN 978-1-4615-9591-5 (eBook)
DOI 10.1007/ 978-1-4615-9591-5

© *1966 Springer Science+Business Media New York*
Originally published by Plenum Press Data Division in 1966
Softcover reprint of the hardcover 1st edition

In the modern organic synthesis industries, one of which is the synthetic rubber industry, ever increasing use is made of physical and physicochemical methods of analysis, which surpass chemical methods in speed, accuracy, and sensitivity. By these methods it is often possible to arrive at the solution of problems in the investigation of complex mixtures of organic products which are not amenable to the usual chemical methods of analysis.

One such physical method is ultraviolet spectrophotometry. The field of application of this method is restricted, in the main, to aromatic compounds and to systems containing double bonds conjugated among themselves or with functional groups. In the synthetic rubber industry ultraviolet spectroscopy finds application in the analysis of a great variety of substances used in that industry: for the determination of impurities in monomers and intermediate products, in the study of the composition of certain polymers, for the quantitative estimation of various ingredients in rubbers, in the control of certain copolymerization processes, and for many other purposes. The method can be used for the identification of certain compounds and can be applied in the determination of the composition of synthetic rubber samples. Shortcomings of the method, which limit its analytical application in certain cases, are the superposition of absorption spectra and their inadequate selectivity.

The atlas gives absorption spectra in the near ultraviolet (200–400 mμ) for 141 different substances met in synthetic rubber manufacture: monomers, polymers, various antioxidants, and other substances used in the synthesis of rubbers (emulsifiers, initiators, regulators, various auxiliary materials, etc.).

All the spectra are represented in the form of graphs of

the relation $a = f(\lambda)$, in which a is the specific extinction coefficient (optical density of a solution containing 1 g of the substance in 1 liter at a layer thickness of 1 cm) and λ is wavelength ($m\mu$). The scales of the spectra vary: they were chosen so as to reveal the characteristics of the spectrum as clearly as possible. The spectra shown in the atlas were determined by the authors in the laboratories of the All-Union Synthetic Rubber Research Institute with an SF-4 spectrophotometer at room temperature (20 ± 3°C). Many of them were determined for the first time. The relative error in the determination is up to 5%, which is usual for the SF-4. In some cases our results do not agree with data in the literature (which are cited in footnotes to the tables). It should be noted that in a number of cases we had to deal with technical products or with substances of unknown purity. This refers mainly to antioxidants and to various ingredients used in the synthesis of rubbers.

Apart from the spectra, at the beginning of each section a table is given which states the solvent in which the measurement was conducted, the wavelength λ_{max} at each absorption maximum, and the specific a and molecular ϵ extinction coefficients at each maximum.

In the legends to the figures we state, successively, the solvent, the concentration c, and the thickness d of the layer of solution.

The atlas does not give the spectra of some aromatic hydrocarbons used in the synthetic rubber industry as solvents or for other purposes (e.g., benzene and toluene) because their spectra can be found in a readily available atlas [1]. The general laws of electronic spectra and the experimental techniques, which have been described in the literature [2, 3], are not expounded here.

The authors hope that, despite possible shortcomings, this atlas will be found useful in the practical work of spectroscopists, not only in the synthetic rubber industry, but in allied fields (plastics, petrochemicals, etc.) in which the compounds whose spectra are given in the atlas can be met.

Contents

CONTENTS

I. MONOMERS

The monomers used in the preparation of synthetic rubbers (Table 1) are substances containing a system of conjugated double bonds or a double bond conjugated with a triple. The absorption spectra of all of them therefore contain strong K bands. When the substance contains a double bond conjugated with a benzene ring, a B band appears in addition to the K band. When the double bond is conjugated with a carbonyl group (acrylic compounds), the K band is considerably weaker.

The character of the absorption spectrum is also affected by the substituents situated at the conjugated bond. When there is a substituent on the second carbon atom of a dienic system, a bathochromic shift in the absorption band is observed (isoprene, chloroprene). Also, substitutents such as methyl or chloro bring about a reduction in the strength of the absorption band. When the methyl group is on the first carbon atom of the dienic system a bathochromic shift in the absorption band is again observed, and there is a still greater reduction in its strength (piperylene). A methyl group on the second carbon atom at a double bond conjugated with a benzene ring brings about only a slight reduction in the strength of the two bands (K and B), but in this case there are hypsochromic shifts in the absorption bands (a-methylstyrene).

When the double bond is conjugated with a pyridine ring, the strengths of the K and B bands are much lower than in the case of conjugation with a benzene ring, and the absorption bands are shifted considerably toward the longer waves.

Since the polymerization of a monomer destroys the conjugated system of bonds, polymers do not give K absorption bands. It is therefore possible to estimate the amount of unchanged monomer from the strength of the absorption band and to follow the progress of the polymerization. Similarly, the residual monomer content of a polymer can be determined spectrophotometrically [4].

1

TABLE 1

SPECTRAL CHARACTERISTICS OF MONOMERS

Figure	Monomer	Solvent	λ_{max}, mμ	α	ϵ
1	1,3-Butadiene	Ethanol	218	570	30,750*
2	Isoprene (2-methyl-1,3-butadiene)	Ethanol	223	316	21,500†
3	Chloroprene (2-chloro-1,3-butadiene)	Ethanol	223	240	21,250
4	Piperylene, 93% of trans form (trans-1,3-penta-diene)	Ethanol	220	250	16,850‡
5	Piperylene, 77.5% of cis form (cis-1,3-penta-diene)	Ethanol	225	235	16,000‡
6	Styrene	Ethanol	208	203	21,100
			248	150	15,500§
		Dioxane	248.5	137	14,300
7	α-Methylstyrene	Ethanol	206	170	20,000
			243	103	12,150
8	Methacrylic acid	Ethanol	207.5	92	7,900
9	Methyl methacrylate (technical)	Ethanol	207.5	84	8,400
10	Butyl methacrylate (technical)	Ethanol	207.5	66	8,300
11	Acrylonitrile	Ethanol	205	90	4,750
12	2-Vinylpyridine	Chloroform	238	109	11,450
			278.5	48	5,050
13	5-Vinyl-2-picoline	Chloroform	244	140	16,650
			284	37	4,400
14	2-Methyl-5-hexen-3-yn-2-ol	50% Ethanol	221.5	138	15,200
			230.5	111	12,200
15	p-(1,1-Dimethyl-4-penten-2-ynyl)phenol	50% Ethanol	223.5	107	19,900
			275.5	10	1,850

*In hexane λ_{max}= 217 mμ, ϵ = 20,900 [5].
†In hexane λ_{max}= 220 mμ, ϵ = 23,900 [5]. In hexane λ_{max} = 220 mμ, ϵ = 21,000 [6].
‡In alcohol λ_{max}= 223.5 mμ, ϵ = 23,000 [8].
§In alcohol λ_{max}= 244 mμ, ϵ = 13,000 [9].

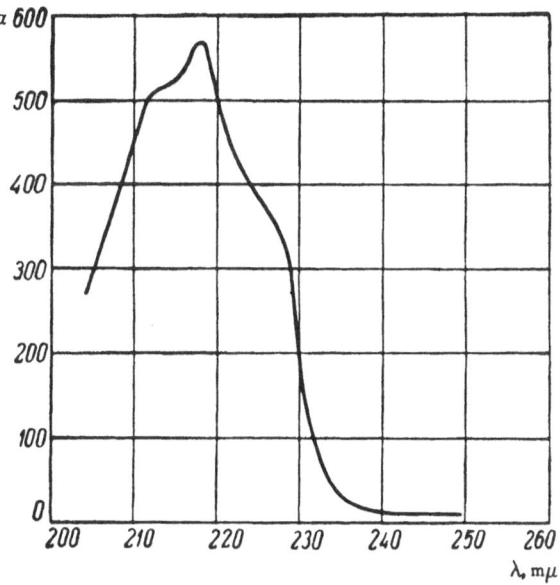

Fig. 1. 1,3-Butadiene

$CH_2 = CH - CH = CH_2$

Ethanol
$C = 0.15$ g/liter
$d = 0.58$ mm

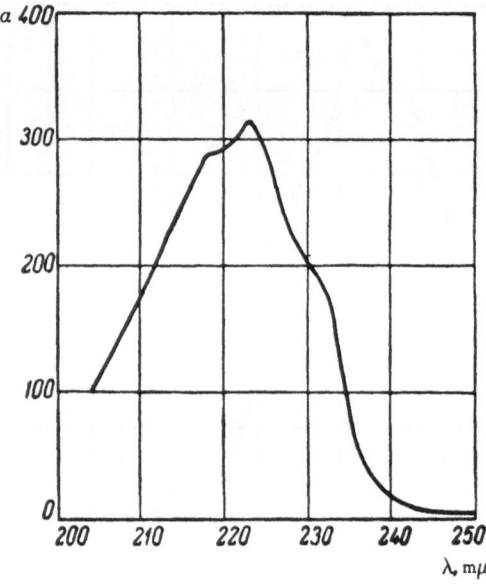

Fig. 2. Isoprene (2-methyl-1,3-butadiene)

$$CH_2 = C - CH = CH_2$$
with CH₃ group:

CH₃
|
CH₂ = C − CH = CH₂

Ethanol
C = 0.15 g/liter
d = 0.58 mm

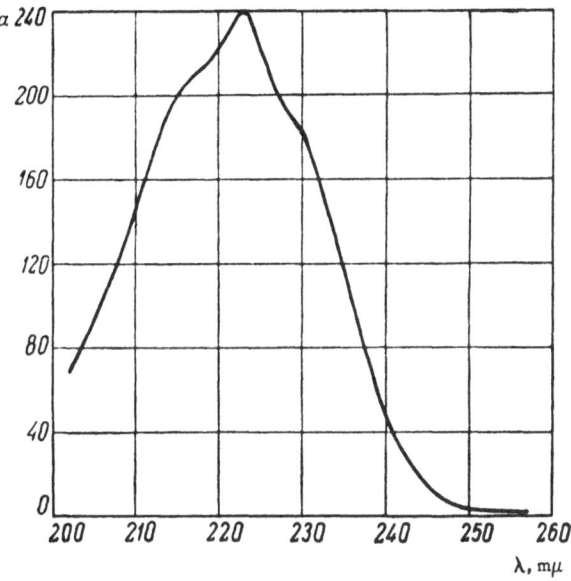

Fig. 3. Chloroprene (2-chloro-1,3-butadiene)

$$CH_2 = \underset{\underset{Cl}{|}}{C} - CH = CH_2$$

Ethanol
$C = 0.82$ g/liter
$d = 0.058$ mm

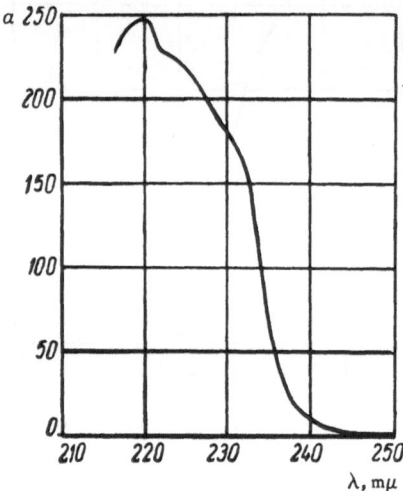

Fig. 4. *trans*-Piperylene (*trans*-1,3-pentadiene)

CH$_2$ = CH – CH = CH – CH$_3$

Ethanol
C = 0.77 g/liter
d = 0.049 mm

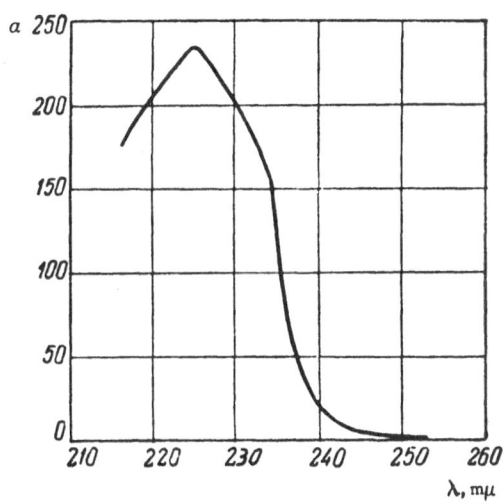

Fig. 5. *cis*-Piperylene (*cis*-1,3-pentadiene)

Ethanol
$C = 0.78$ g/liter
$d = 0.049$ mm

$$CH_2 = CH - CH = CH - CH_3$$

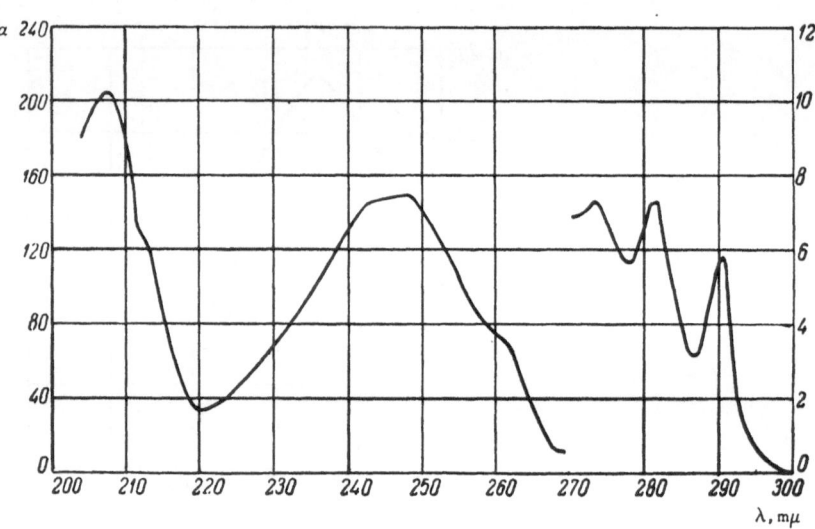

Fig. 6. Styrene

C₆H₅ CH = CH₂

Ethanol
$C = 1.06$ g/liter
$d = 0.058$ and 1.004 mm

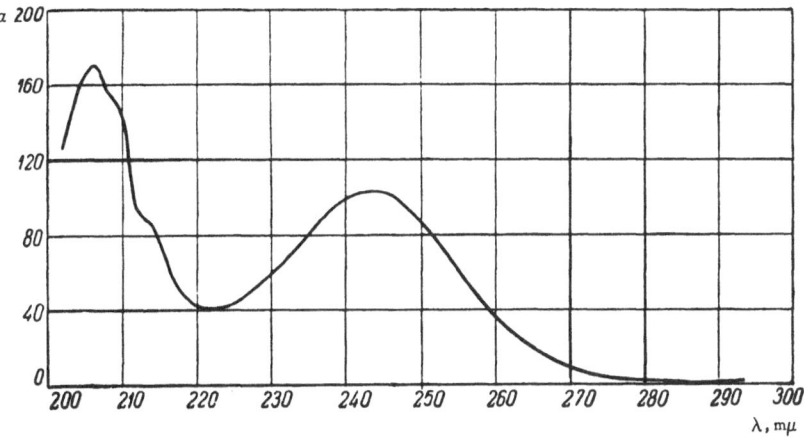

Fig. 7. α-Methylstyrene

$$\begin{array}{c} CH_3 \\ | \\ C_6H_5\,C = CH_2 \end{array}$$

Ethanol
$C = 1.016$ g/liter
$d = 0.058$ mm

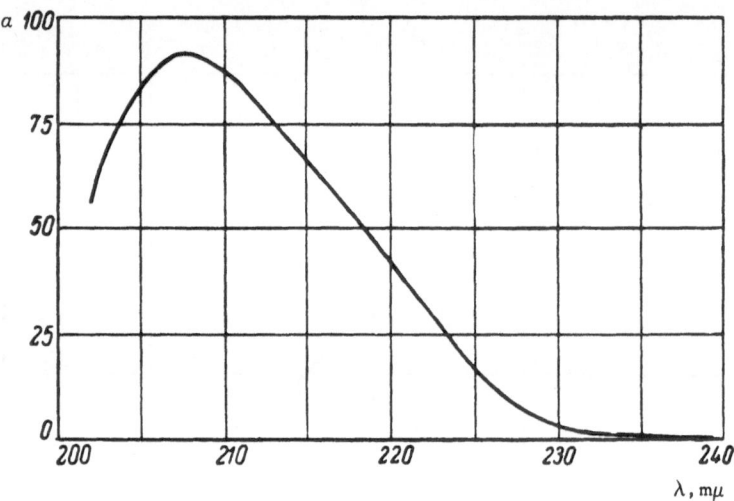

Fig. 8. Methacrylic acid

$$CH_2 = \overset{\overset{\displaystyle CH}{|}}{C} - C \overset{\displaystyle O}{\underset{\displaystyle OH}{}}$$

Ethanol
$C = 1.01$ g/liter
$d = 0.112$ mm

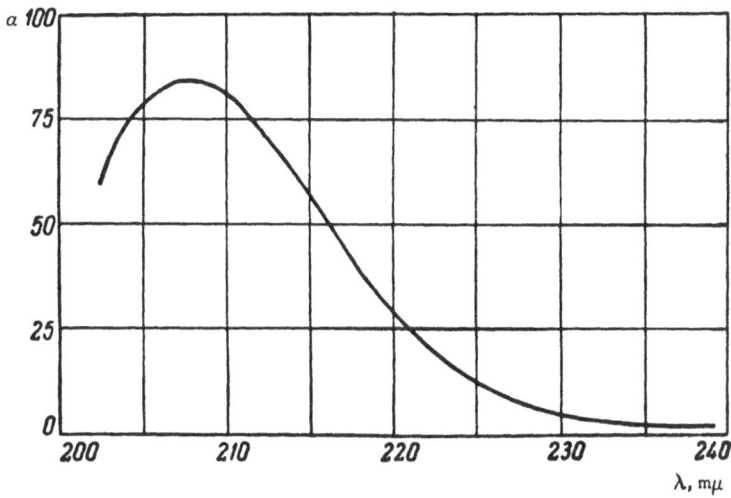

Fig. 9. Methyl methacrylate

$$CH_2 = C - C\begin{smallmatrix} \\ \end{smallmatrix}$$

Ethanol
$C = 1.96$ g/liter
$d = 0.058$ mm

11

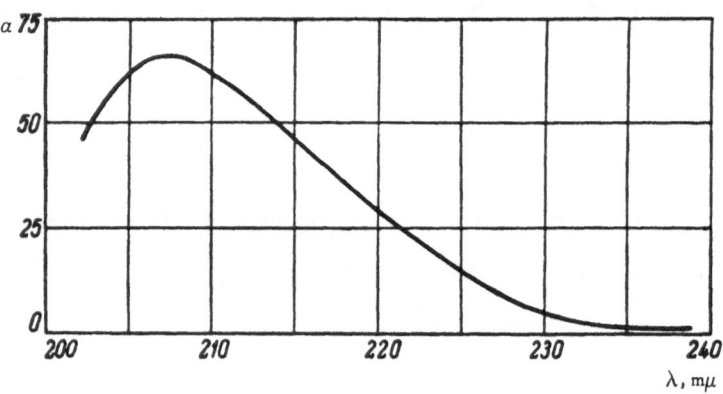

Fig. 10. Butyl methacrylate

$$CH_2= \overset{\overset{\displaystyle CH_3}{\displaystyle |}}{C} - C \overset{\displaystyle \nearrow O}{\searrow_{OC_4H_9}}$$ Ethanol
$C = 2.64$ g/liter
$d = 0.058$ mm

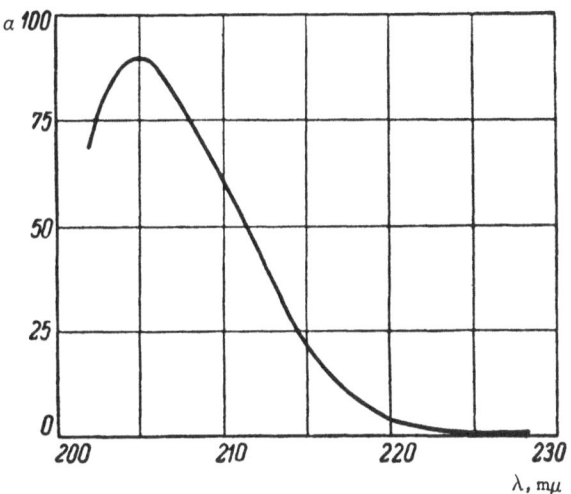

Fig. 11. Acrylonitrile

$$CH_2= CH-C \equiv N$$

Ethanol
$C = 1.79$ g/liter
$d = 0.058$ mm

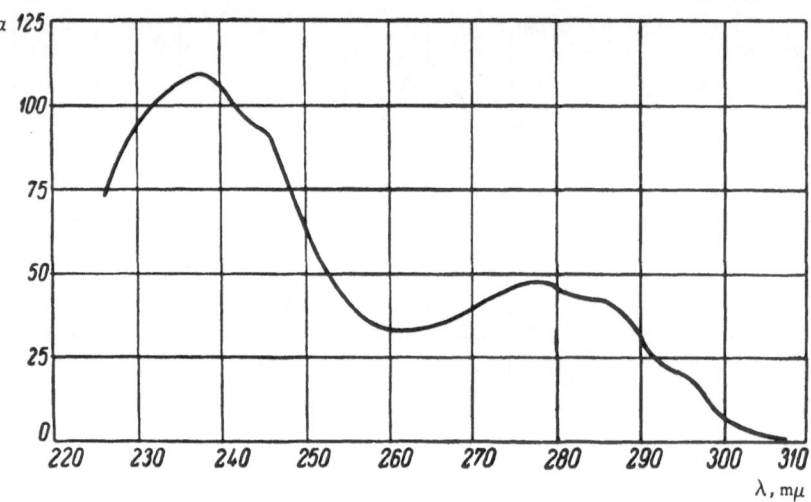

Fig. 12. 2-Vinylpyridine

$CH_2 = CH - C_5 H_4 N$

Chloroform
$C = 1.00$ g/liter
$d = 0.058$ mm

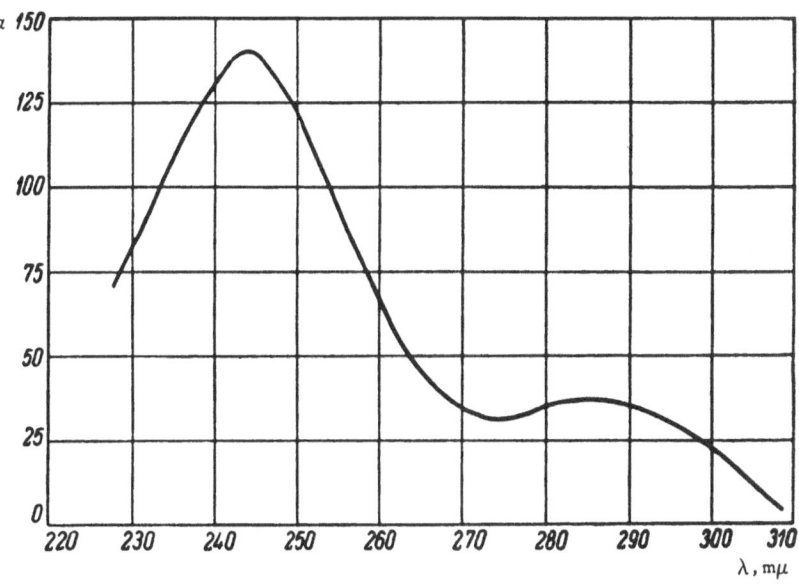

Fig. 13. 5-Vinyl-2-picoline

$CH_3C_5H_3NCH = CH_2$

Chloroform
$C = 0.10$ g/liter
$d = 0.506$ mm

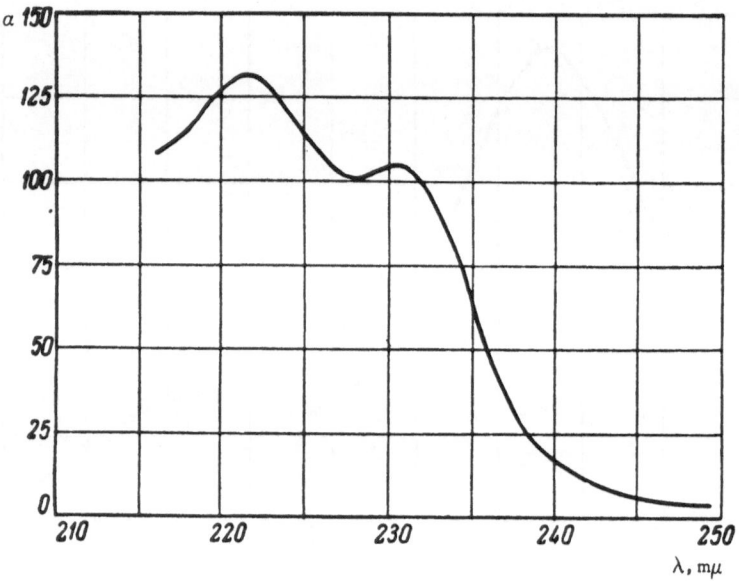

Fig. 14. 2-Methyl-5-hexen-3-yn-2-ol

$$CII_2 = CII \quad C \equiv C \quad \underset{\underset{CH_3}{|}}{\overset{\overset{CH_3}{|}}{C}} \cdots OH$$

50% ethanol
$C = 1.00$ g/liter
$d = 0.058$ mm

16

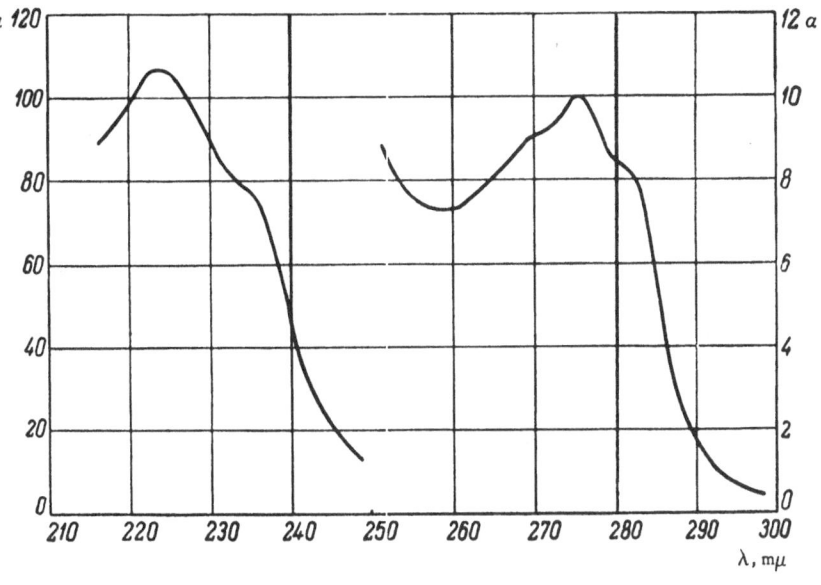

Fig. 15. p-(1,1-Dimethyl-4-penten-2-ynyl) phenol

$$CH_2 = CH - C \equiv C - \underset{\underset{CH_3}{|}}{\overset{\overset{CH_3}{|}}{C}} - C_6H_4OH$$

50% ethanol
$C = 1.00$ g/liter
$d = 0.058$ and 0.506 mm

II. POLYMERS

In the polymerization of monomers containing conjugated systems of bonds, the latter are destroyed and the associated K absorption bands disappear. Usually, therefore, polymers (Table 2) do not have absorption bands in the near ultraviolet. However, in the polymerization or copolymerization of monomers containing a benzene or pyridine ring in the molecule the B band associated with this ring does not disappear in the spectrum of the polymer. Hence, whereas the study of the polymerization processes of monomers not containing benzene or pyridine rings can be conducted only by determining the content of unchanged monomer, when there are B bands it is possible to determine directly the amount of monomer that has polymerized [10].

It is interesting to note the difference in the spectra of polystyrenes prepared by emulsion polymerization and by catalytic polymerization in solution both in the position and in the strength of the absorption bands. This points to a difference in structure between the polymers. The character of the absorption bands is preserved even in copolymerization with other monomers, depending on the method of polymerization.

The absorption bands in the spectra of thiokols arise from the sulfide bonds; they are R bands.

TABLE 2

SPECTRAL CHARACTERISTICS OF POLYMERS

Figure	Structural unit	Solvent	λ_{max}, mμ	α	ϵ
16	Styrene, in emulsion copolymers	Chloroform	255.7	1.85	190
			262.5	2.20	230
			269.5	1.60	165
17	Styrene, in products of catalytic copolymerization in solution	Chloroform	254.5	7.05	730
18	(3-p-Hydroxyphenyl-3-methyl-1-butynyl) ethylene in copolymers with 1,3-butadiene	Chloroform	276	8.0	1500
			282.5	6.9	1300
19	(2-Pyridyl) ethylene in copolymers with 1,3-butadiene	Chloroform	258	32	3350
			263	33.5	3500
			270	25	2600
20	(5-Methyl-2-pyridyl) ethylene in copolymers with 1,3-butadiene	Chloroform	269.5	28	2500
			276	21.5	2550
21	Unit of thiokol T-1	Dioxane	210	5.00	830
			250	2.30	380
22	Unit of thiokol T-2	Dioxane	208	5.30	1350
			250	1.85	470
23	Unit of thiokol T-3	Dioxane	207	3.90	1400
			249	1.20	435

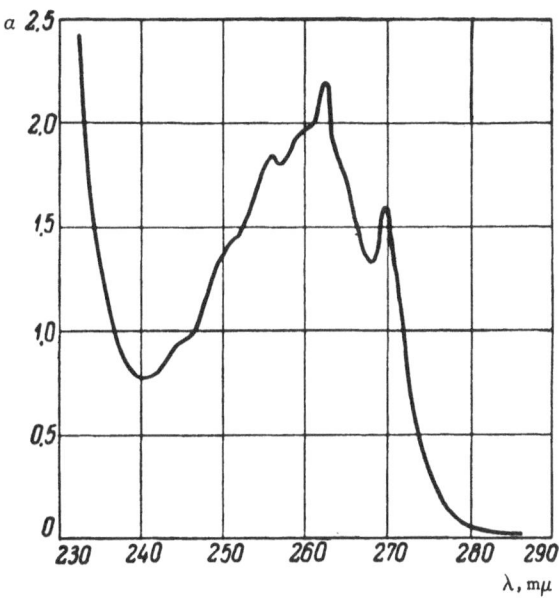

Fig. 16. Styrene unit in emulsion copolymers

$$\begin{bmatrix} -CH-CH_2- \\ | \\ C_6H_5 \end{bmatrix}$$

Chloroform
$C = 10.0$ g/liter
$d = 0.506$ mm

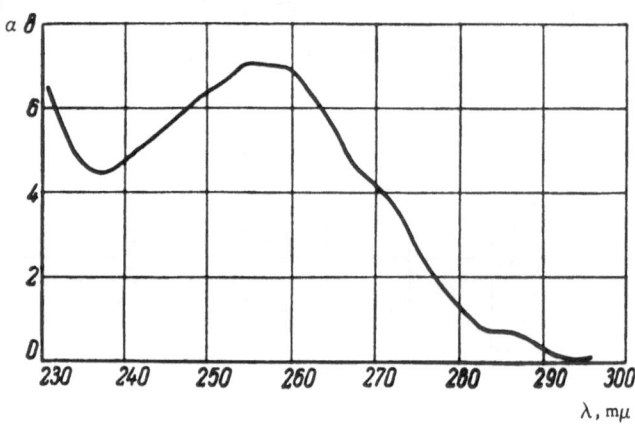

Fig. 17. Styrene unit in products of catalytic copolymerization in solution

$$\begin{bmatrix} -CH=CH_2- \\ | \\ C_6H_5 \end{bmatrix}$$

Chloroform
$C = 0.94$ g/liter
$d = 1.012$ mm

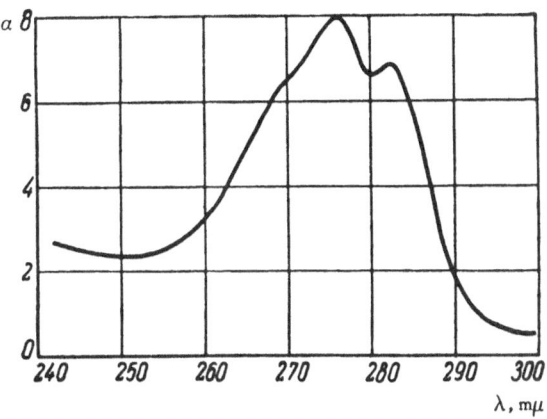

Fig. 18. (3-*p*-Hydroxyphenyl-3-methyl-1-butynyl) ethylene
unit in copolymers with 1,3-butadiene

$$\begin{bmatrix} -CH-CH_2- \\ \quad | \\ C \equiv C - C(CH_3)_2C_6H_4OH \end{bmatrix}$$

Chloroform
$C = 1.00$ g/liter
$d = 1.012$ mm

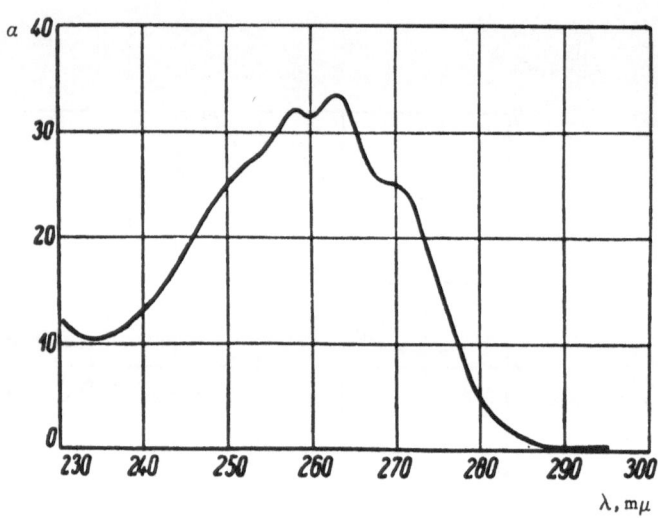

Fig. 19. (2-Pyridyl) ethylene unit in copolymers with 1,3-butadiene

$$\begin{bmatrix} -CH-CH_2- \\ | \\ C_5H_4N \end{bmatrix}$$

Chloroform
$C = 0.96$ g/liter
$d = 0.210$ mm

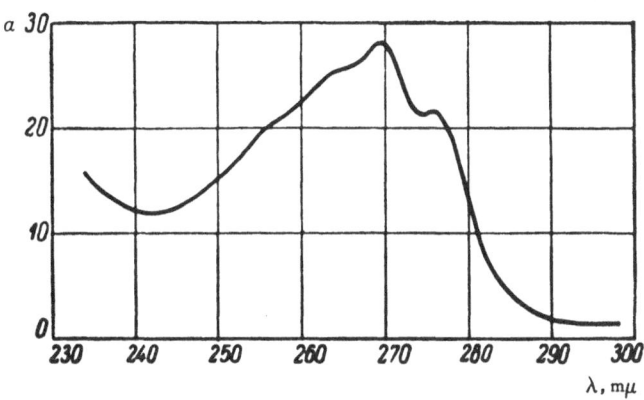

Fig. 20. (5-Methyl-2-pyridyl) ethylene unit in copolymers with 1,3-butadiene

$$\left[\begin{array}{c} -CH-CH_2- \\ | \\ C_5 H_3 NCH_3 \end{array} \right]$$

Chloroform
$C = 1.20$ g/liter
$d = 0.210$ mm

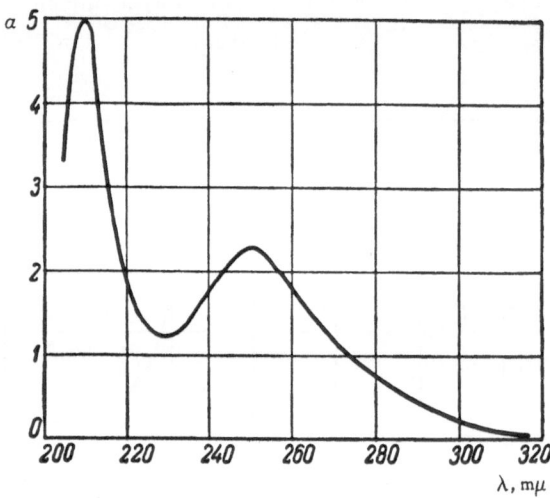

Fig. 21. Unit of thiokol T-1

$[-S(CH_2)_2OCH_2O(CH_2)_2S-]$

Dioxane
$C = 10.0$ g/liter
$d = 0.210$ mm

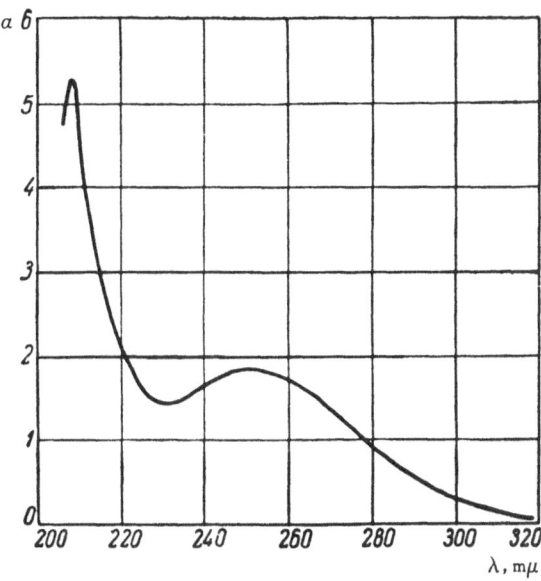

Fig. 22. Unit of thiokol T-2

$$[-S(CH_2)_2O(CH_2)_2OCH_2O(CH_2)_2O(CH_2)_2S-]$$

Dioxane
$C = 10.0$ g/liter
$d = 0.210$ mm

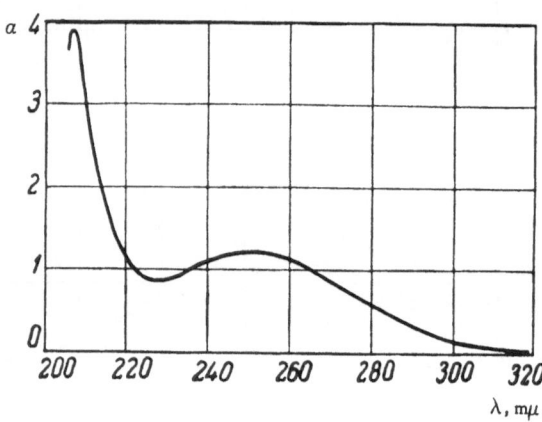

Fig. 23. Unit of thiokol T-3

Dioxane
C = 10.0 g/liter
d = 0.210 mm

$[-S(CH_2)_2O(CH_2)_2O(CH_2)_2OCH_2O(CH_2)_2O(CH_2)_2O(CH_2)_2S-]$

III. ORGANOSILICON COMPOUNDS

The application of ultraviolet spectroscopy in the investigation of organosilicon compounds is limited in the main to those containing aromatic rings. However, the extensive application of cocondensation reactions to give copolymers containing dimethylsiloxane units and a great variety of aromatic siloxane units opens up wide possibilities for the use of the absorption spectra of the latter, both for the purpose of identifying these aromatic derivatives and for their quantitative determination in copolymers [11, 12].

Nearly all the organosilicon compounds whose spectra are given here (Table 3) are polysiloxanes. Their segregation in a special section is justified by the specific structure of the polymer chain. Also, the spectra of a few monomeric compounds are given along with those of the corresponding polysiloxanes, which makes it possible to trace changes in the spectrum due to the presence of various substituents on silicon. In view of the fact that almost all of the polyarylsiloxanes given here are copolymers containing dimethylsiloxane units, extinction coefficients were calculated in terms of one arylsiloxane elementary unit.

The spectra of these compounds are essentially the spectra of benzene or biphenyl containing one or more silicon substituents, and they contain B bands. The positions and intensities of the absorption maxima depend on the number and structures of the substituents in the benzene ring. The introduction of CH_2 groups between the silicon atom and the aromatic ring also has an effect. It is interesting that in this case a certain alternation of properties is observed, as is found in some homologous series: the absorption coefficients of compounds containing an odd number of CH_2 groups between silicon and phenyl are higher than those of adjacent compounds with an even number of CH_2 groups [11, 13]. Moreover, the introduction of CH_2 groups causes a smoothing-out of the spectrum, as also does the introduction of a second substitutent into the phenyl group.

Some of the organosilicon compounds whose spectra are given here were synthesized at the All-Union Synthetic Rubber Research Institute for the first time.

TABLE 3

SPECTRAL CHARACTERISTICS OF ORGANOSILICON STRUCTURAL UNITS IN COPOLYMERS CONTAINING DIMETHYLSILOXANE UNITS

Figure	Structural unit (or compound)	Solvent	λ_{max}, mμ	a	ϵ
24	Methylphenylsiloxane unit	Chloroform	253.5	1.45	197
			259.0	2.10	286
			264.0	2.25	306
			270.5	1.70	231
25	(Chloromethyl) phenylsilox-ane unit	Chloroform	254.0	1.30	222
			259.5	1.90	324
			264.5	2.30	392
			271.0	1.85	315
26	Ethylphenylsiloxane unit	Chloroform	253.5	1.40	210
			259.0	2.00	300
			264.0	2.20	330
			270.5	1.60	240
27	Phenylvinylsiloxane unit	Chloroform	253.5	1.45	215
			259.0	2.10	311
			264.0	2.25	353
			270.5	1.70	251
28	Methyl (m-trifluoromethyl-phenyl) siloxane unit	Chloroform	259.0	1.80	367
			264.5	2.30	469
			271.0	1.75	357
29	Benzylmethylsiloxane unit	Chloroform	261.0	2.25	337
			267.0	2.80	420
			274.0	2.40	360
30	(a-Methylbenzyl) methylsil-oxane unit	Chloroform	261.0	2.00	328
			266.5	2.25	369
			273.5	1.65	270
31	Methylphenethylsiloxane unit	Chloroform	262.5	1.65	270
			270.0	1.30	213
32	Methyl (3-phenylpropyl)-siloxane unit	Chloroform	261.5	1.75	308
			269.0	1.35	238

Figure	Structural unit (or compound)	Solvent	λ_{max}, mμ	a	ϵ
33	Methyl (4-phenylbutyl)-siloxane unit	Chloroform	262.0 269.0	1.55 1.20	291 226
34	(Anilinomethyl) methyl-siloxane unit	Chloroform	248.0 297.0	80 15	13,200 2,475
35	Methyl-1-naphthylsilox-ane unit	Chloroform	273.5 283.0 293.0	36 42 30	6,750 7,850 5,600
36	Diphenylsilanediol	Chloroform	259.0 264.0 270.0	3.00 3.05 2.30	648 659 497
37	Diphenylsiloxane unit	Chloroform	260.0 265.5 271.5	3.25 3.65 2.75	643 723 545
38	(Methylphenylsilylene)-(ethylene) (methyl-phenylsiloxane) unit	Chloroform	254.0 259.5 264.5 270.5	1.70 2.20 2.10 1.40	483 625 596 398
39	p-Phenylenebis [dimethyl-silane]	Chloroform	265.0 270.0 276.5	2.10 2.25 1.70	407 437 330
40	p-Phenylenebis [dimethyl-silanol]	Chloroform	264.5 269.5 275.5	1.20 1.45 1.25	271 328 283
41	(Dimethylsilylene) (p-phenylene) (dimethyl-siloxane) unit	Chloroform	264.5 269.5 275.5	2.10 2.35 2.00	437 489 716
42	[Methyl (3,3,3-trifluro-propyl) silylene] (p-phenylene) [methyl-(3,3,3-trifluoropropyl) siloxane] unit	Chloroform	270.0 276.5	1.60 1.40	595 521
43	(Dimethylsilylene) (m-phenylene) (dimethyl-siloxane) unit	Chloroform	264.0 269.0 275.0	1.60 1.70 1.20	334 354 250
44	4,4'-Biphenylylenebis-[dimethylsilane]	Chloroform	265.0	9.50	2,550
45	4,4'-Biphenylylenebis-[chlorodimethylsilane]	Chloroform	260.0	4.75	1,700
46	(Oxydi-p-phenylene)-bis[dimethylsilanol]	Chloroform	218.0	57	18,100
47	(Dimethylsilylene)-p-phenylene-oxy-p-phenyl-ene (dimethylsiloxane) unit	Chloroform	218.5 239.0	62.5 57.5	18,750 17,250

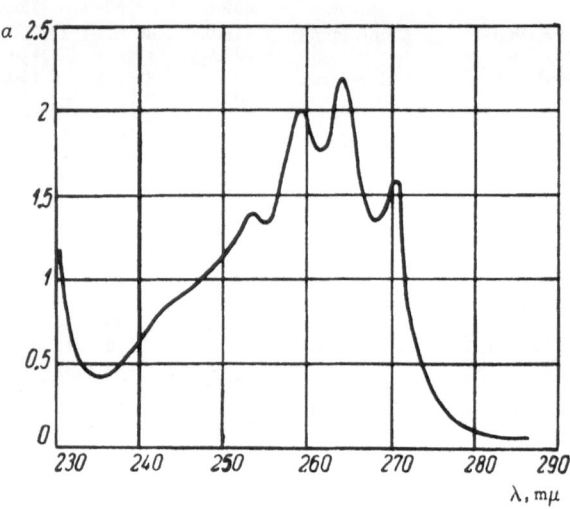

Fig. 24. Methylphenylsiloxane unit

$$\begin{bmatrix} & C_6H_5 & \\ & | & \\ -Si & - O - \\ & | & \\ & CH_3 & \end{bmatrix}$$

Chloroform
$C = 10.0$ g/liter
$d = 0.205$ mm

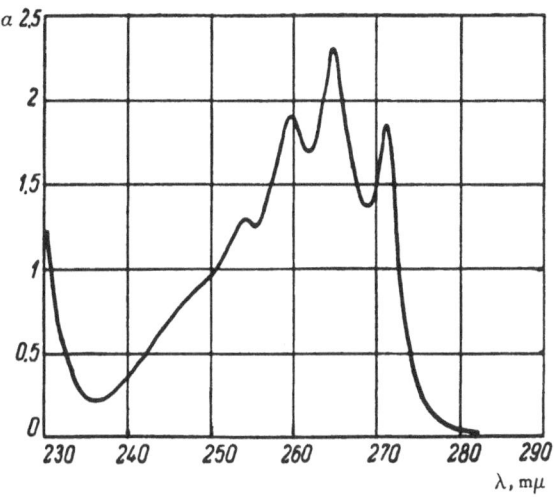

Fig. 25. (Chloromethyl) phenylsiloxane unit

$$\begin{bmatrix} C_6H_5 \\ | \\ -Si-O- \\ | \\ CH_2Cl \end{bmatrix}$$

Chloroform
$C = 10.2$ g/liter
$d = 0.205$ mm

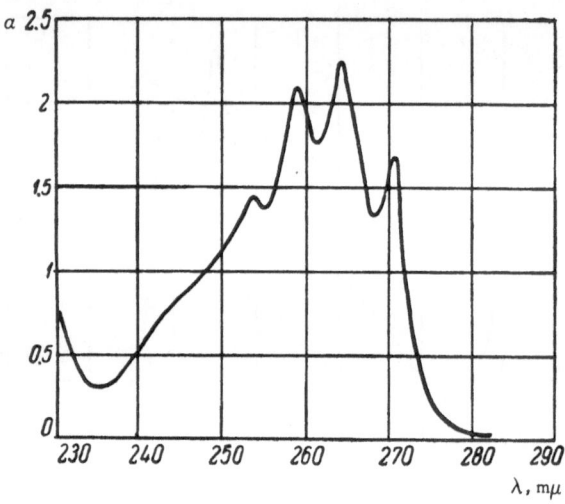

Fig. 26. Ethylphenylsiloxane unit

$$\begin{bmatrix} & C_6H_5 & \\ & | & \\ -Si & -O- \\ & | & \\ & C_2H_5 & \end{bmatrix}$$

Chloroform
$C = 9.25$ g/liter
$d = 0.205$ mm

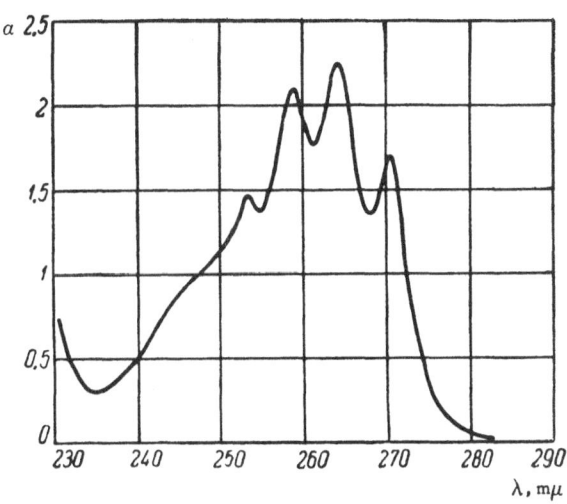

Fig. 27. Phenylvinylsiloxane unit

$$\begin{bmatrix} C_6H_5 \\ | \\ -Si-O- \\ | \\ CH=CH_2 \end{bmatrix}$$

Chloroform
$C = 10.0$ g/liter
$d = 0.205$ mm

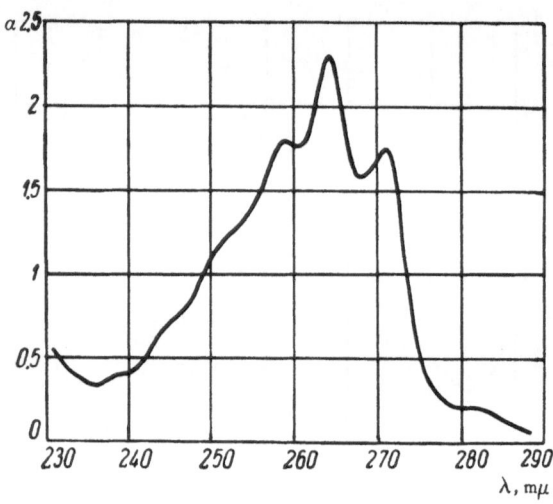

Fig. 28. Methyl (*m*-trifluoromethylphenyl) siloxane unit

$$\begin{bmatrix} \begin{array}{c} C_6H_4CF_3 \\ | \\ Si \quad O \\ | \\ CH_3 \end{array} \end{bmatrix}$$

Chloroform
$C = 50.0$ g/liter
$d = 0.107$ mm

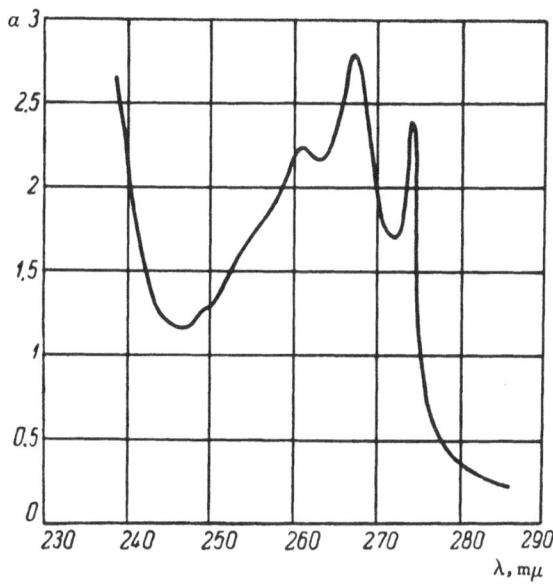

Fig. 29. Benzylmethylsiloxane unit

$$\begin{bmatrix} & CH_2C_6H_5 \\ & | \\ -Si&-O- \\ & | \\ & CH_3 \end{bmatrix}$$

Chloroform
$C = 9.25$ g/liter
$d = 0.205$ mm

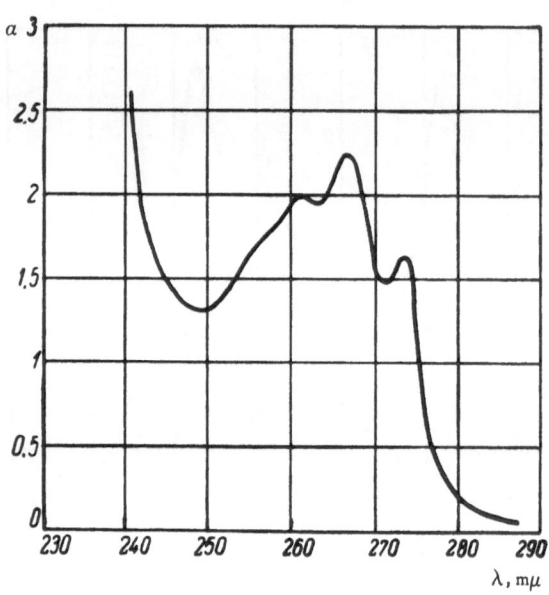

Fig. 30. (a-Methylbenzyl) methylsiloxane unit

$$\begin{bmatrix} CH_3 \\ | \\ CH-C_6H_5 \\ | \\ -Si-O- \\ | \\ CH_3 \end{bmatrix}$$

Chloroform
$C = 9.95$ g/liter
$d = 0.205$ mm

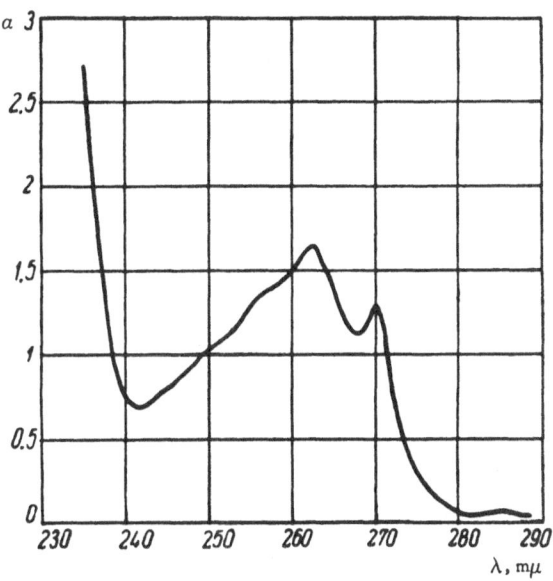

Fig. 31. Methylphenethylsiloxane unit

$$\begin{bmatrix} CH_2CH_2C_6H_5 \\ | \\ -Si-O- \\ | \\ CH_3 \end{bmatrix}$$

Chloroform
$C = 9.95$ g/liter
$d = 0.510$ mm

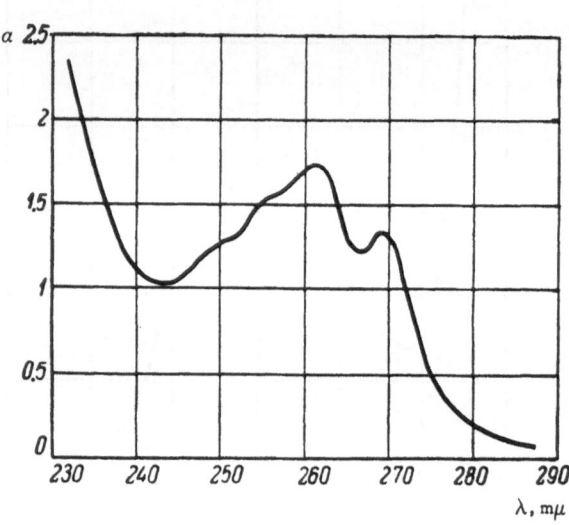

Fig. 32. Methyl (3-phenylpropyl) siloxane unit

$$
\left[\begin{array}{c} CH_2CH_2CH_2C_6H_5 \\ | \\ -Si-O- \\ | \\ CH_3 \end{array} \right]
$$

Chloroform
$C = 10.6$ g/liter
$d = 0.205$ mm

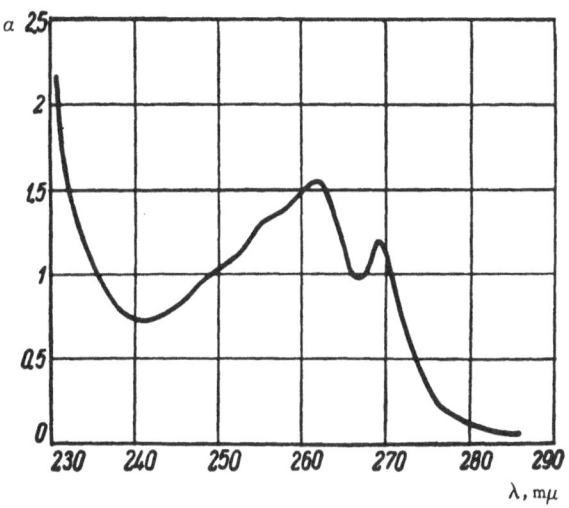

Fig. 33. Methyl(4-phenylbutyl) siloxane unit

$$\begin{bmatrix} CH_2CH_2\ CH_2CH_2C_6H_5 \\ | \\ -Si-O- \\ | \\ CH_3 \end{bmatrix}$$

Chloroform
$C = 11.3$ g/liter
$d = 0.205$ mm

41

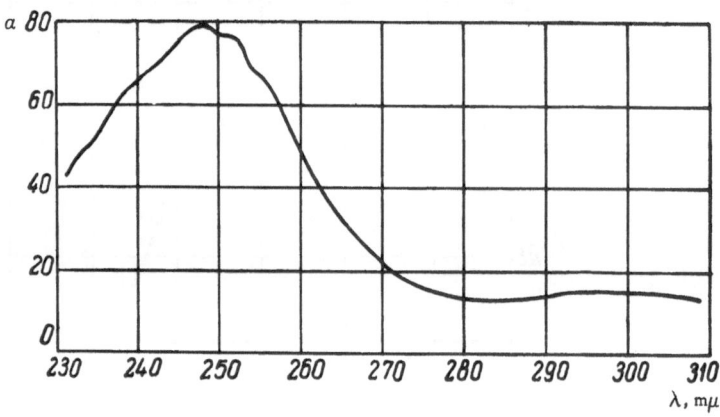

Fig. 34. (Anilinomethyl) methylsiloxane unit

$$\begin{bmatrix} CH_2NHC_6H_5 \\ | \\ -Si-O- \\ | \\ CH_3 \end{bmatrix}$$

Chloroform
$C = 1.05$ g/liter
$d = 0.107$ mm

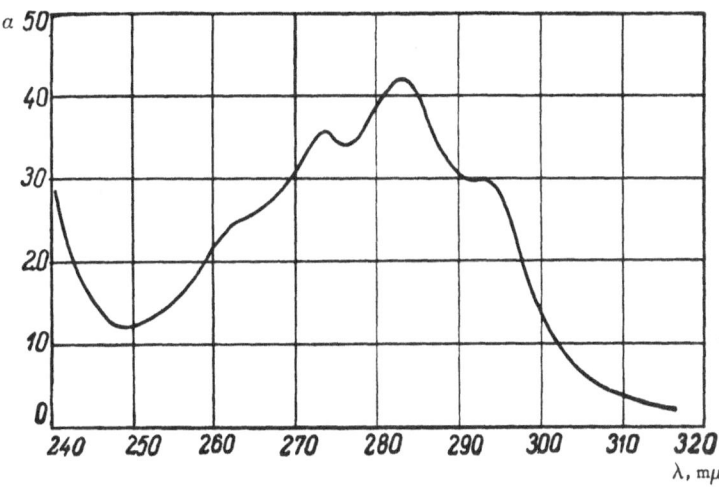

Fig. 35. Methyl-1-naphthylsiloxane unit

$$\begin{bmatrix} C_{10}H_7 \\ | \\ -Si-O- \\ | \\ CH_3 \end{bmatrix}$$

Chloroform
$C = 5.84$ g/liter
$d = 0.066$ mm

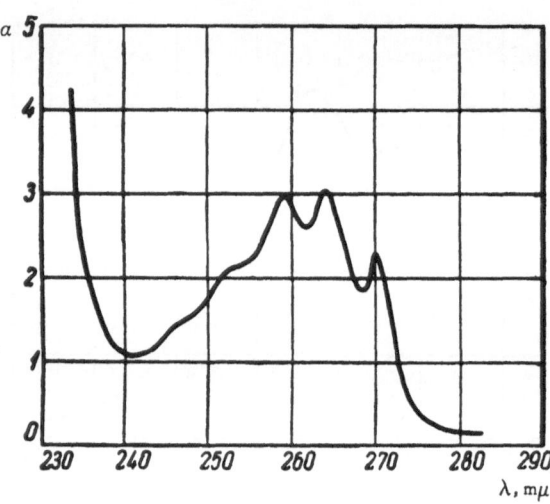

Fig. 36. Diphenylsilanediol

$$
\begin{array}{l}
\quad\ \ C_6H_5 \\
\quad\ \ | \\
H\overset{\cdot}{O}-\overset{\cdot}{S}i-OH \\
\quad\ \ | \\
\quad\ \ C_6H_5
\end{array}
$$

Chloroform
$C = 10.0$ g/liter
$d = 0.212$ mm

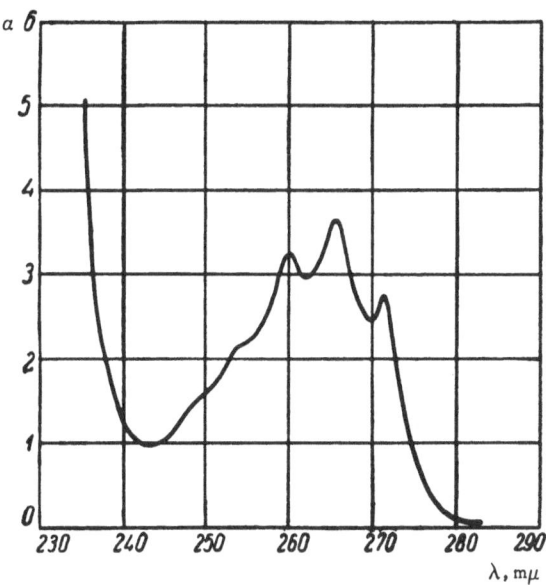

Fig. 37. Diphenylsiloxane unit

$$\left[\begin{array}{c} C_6H_5 \\ | \\ -Si-O- \\ | \\ C_6H_5 \end{array}\right]$$

Chloroform
$C = 10.0$ g/liter
$d = 0.212$ mm

45

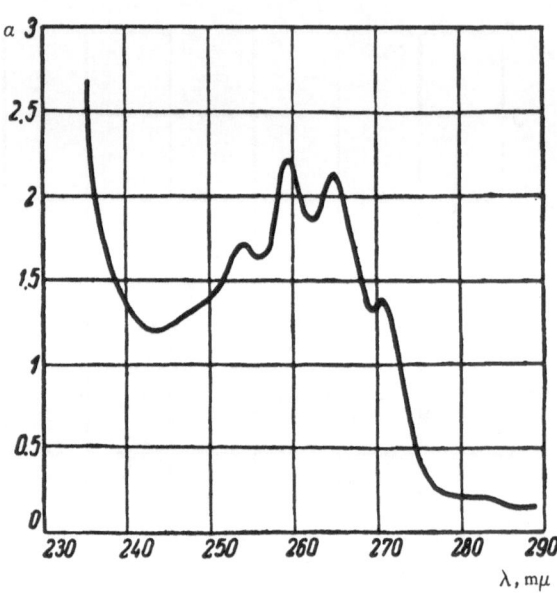

Fig. 38. (Methylphenylsilylene) (ethylene) (methylphenylsiloxane) unit

$$\begin{bmatrix} & \overset{\displaystyle C_6H_5}{\underset{\displaystyle CH_3}{\overset{|}{\underset{|}{-Si}}}} -CH_2-CH_2-\overset{\displaystyle C_6H_5}{\underset{\displaystyle CH_3}{\overset{|}{\underset{|}{Si}}}} -O & \end{bmatrix}$$

Chloroform
C = 8.40 g/liter
d = 0.112 mm

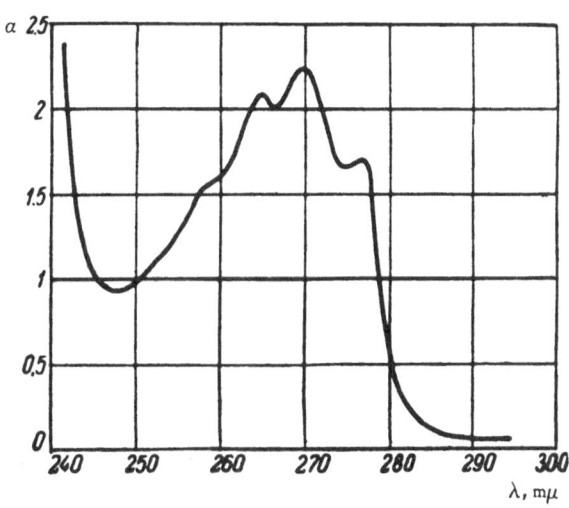

Fig. 39. p-Phenylenebis [dimethylsilane]

$$
\begin{array}{cc}
CH_3 & CH_3 \\
| & | \\
H-Si-C_6H_4-Si-H \\
| & | \\
CH_3 & CH_3
\end{array}
$$

Chloroform
$C = 10.4$ g/liter
$d = 0.506$ mm

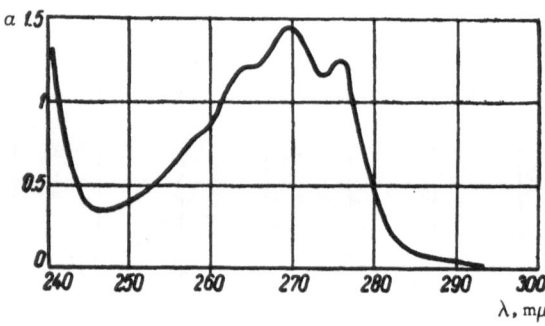

Fig. 40. *p*-Phenylenebis [dimethylsilanol]

$$
\begin{array}{ccc}
& CH_3 & CH_3 \\
& | & | \\
HO-Si-C_6H_4-Si-OH & \\
& | & | \\
& CH_3 & CH_3
\end{array}
$$

Chloroform
$C = 5.11$ g/liter
$d = 0.506$ mm

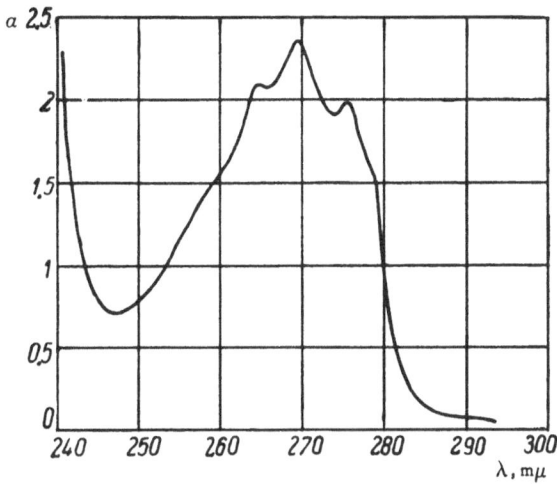

Fig. 41. (Dimethylsilylene) (p-phenylene) (dimethylsiloxane) unit

$$\begin{bmatrix} & CH_3 & & CH_3 \\ & | & & | \\ -Si & - C_6H_4 - & Si & -O- \\ & | & & | \\ & CH_3 & & CH_3 \end{bmatrix}$$

Chloroform
$C = 10.43$ g/liter
$d = 0.212$ mm

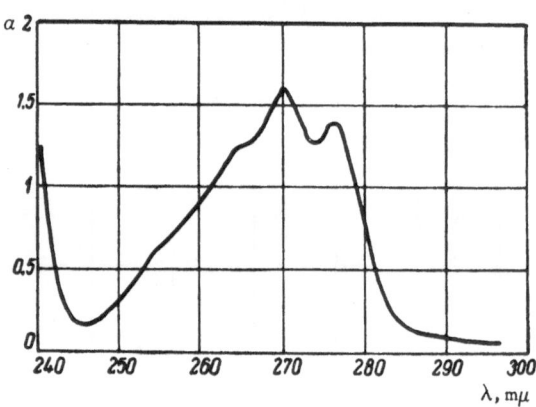

Fig. 42. [Methyl(3,3,3-trifluoropropyl) silylene] (*p*-phenylene) [methyl(3,3,3-trifluoropropyl) siloxane] unit

$$\begin{bmatrix} (CH_2)_2CF_3 & CH_3 \\ | & | \\ -Si-C_6H_4-Si-O- \\ | & | \\ CH_3 & (CH_2)_2CF_3 \end{bmatrix}$$

Chloroform
$C = 10.03$ g/liter
$d = 0.506$ mm

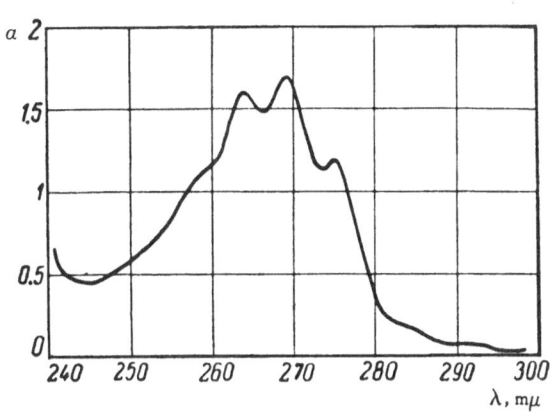

Fig. 43. (Dimethylsilylene) (*m*-phenylene) (dimethylsiloxane) unit

$$\begin{bmatrix} & \overset{\displaystyle CH_3}{\underset{\displaystyle CH_3}{|}} & & \overset{\displaystyle CH_3}{\underset{\displaystyle CH_3}{|}} \\ -Si & -C_6H_4 - & Si - O - \\ & | & & | \end{bmatrix}$$

Chloroform
$C = 10.86$ g/liter
$d = 0.506$ mm

51

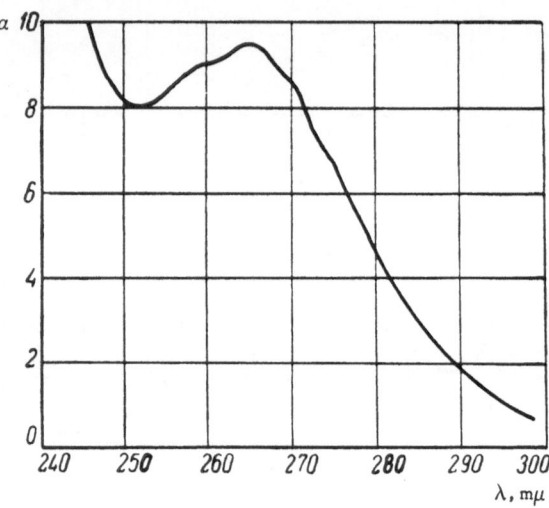

Fig. 44. 4,4′-Biphenylylenebis [dimethylsilane]

$$
\begin{array}{ccc}
CH_3 & & CH_3 \\
| & & | \\
H-Si-C_6H_4C_6H_4-Si-H \\
| & & | \\
CH_3 & & CH_3
\end{array}
$$

Chloroform
$C = 11.4$ g/liter
$d = 0.212$ mm

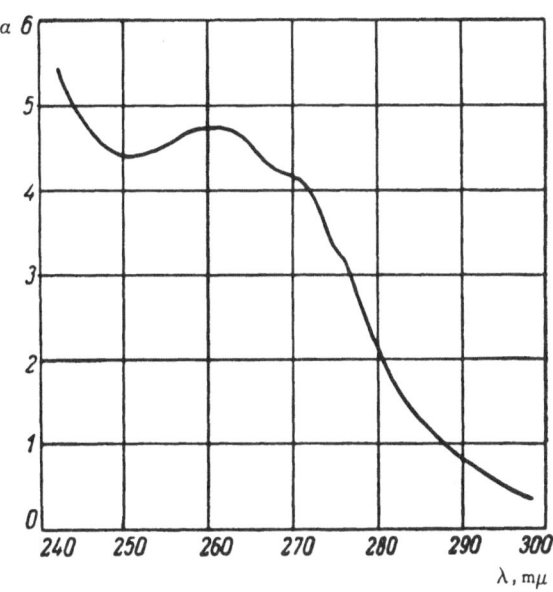

Fig. 45. 4,4′-Biphenylylenebis [chlorodimethylsilane]

$$CH_3 \qquad CH_3$$
$$| \qquad\qquad |$$
$$Cl - Si - C_6H_4C_6H_4 - Si - Cl$$
$$| \qquad\qquad |$$
$$CH_3 \qquad CH_3$$

Chloroform
$C = 12.4$ g/liter
$d = 0.112$ mm

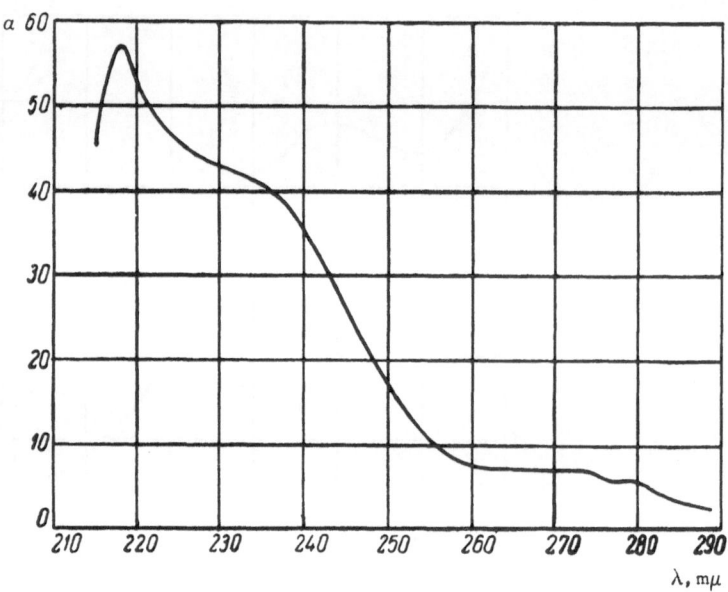

Fig. 46. (Oxydi-p-phenylene) bis [dimethylsilanol]

$$HO-\underset{\underset{CH_3}{|}}{\overset{\overset{CH_3}{|}}{Si}}-C_6H_4OC_6H_4-\underset{\underset{CH_3}{|}}{\overset{\overset{CH_3}{|}}{Si}}-OH$$

Chloroform
$C = 3.06$ g/liter
$d = 0.058$ mm

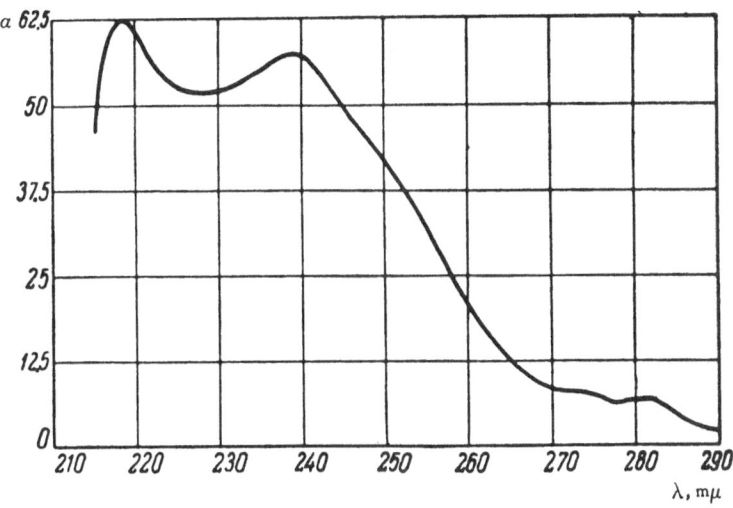

Fig. 47. (Dimethylsilylene)-p-phenylene-oxy-p-phenylene (dimethylsiloxane) unit

$$\left[\begin{array}{ccc} CH_3 & & CH_3 \\ | & & | \\ -Si-C_6H_4OC_6H_4- & Si-O- \\ | & & | \\ CH_3 & & CH_3 \end{array}\right]$$

Chloroform
$C = 3.05$ g/liter
$d = 0.058$ mm

IV. ANTIOXIDANTS

The aging of polymers, which is manifested by the worsening of the physicomechanical, dielectric, optical, and other properties of polymeric materials during storage, treatment, or use, is the result of the occurrence in these materials of various processes induced by heat, light, atmospheric oxygen and ozone, and other external factors.

Many of the antioxidants used for the protection of rubbers from thermal and oxidative aging possess a combination of various properties and are used for the protection of numerous polymeric materials from aging by light, ozone, and other agencies. For this reason the material in this section will be of interest to a large number of investigators working on the stabilization of high-polymer materials.

In this section we present the ultraviolet spectra of the most widely used antioxidants produced in the Soviet Union and abroad. We have also determined and published the spectra of products not yet in commercial production. This applies mainly to p-phenylenediamine and p-anisidine derivatives and to phosphorous esters.

We have classified antioxidants as follows in accordance with their chemical structures:

1) secondary amines (Table 4)
2) p-anisidine derivatives (Table 5);
3) p-phenylenediamine derivatives (Table 6);
4) phenol derivatives (Table 7);
5) phosphorous esters (Table 8);
6) hydroquinone derivatives (Table 9);
7) sulfides (Table 10);
8) hydroquinoline derivatives (Table 11);
9) various antioxidants (Table 12).

From our point of view this is the most convenient classification, for it allows the antioxidants to be grouped not only in accordance with the form of the spectral curve, but usually also in accordance with the mechanism of the inhibiting action on the oxidation process in rubber.

In the main, these spectra were determined for technical products whose purities we did not establish. The data obtained can be used successfully for the identification and quantitative determination of antioxidants in high-polymer materials.

In view of the fact that all antioxidants are aromatic compounds, their spectra contain not only K bands, but also B bands. The position of the bands and the form of the spectral curves are determined by the various peculiarities of structure and the nature of the substituents in the aromatic rings. In these spectra the relation of the character of the spectrum to the structure of the molecule shows up most clearly.

We must point out the peculiar properties of the spectra of antioxidants containing free hydroxy groups (phenol and hydroquinone derivatives and thiodiphenols) and of phosphorous esters. These show as changes in the spectrum in alcoholic alkaline solution, and the changes depend on the structure of the antioxidant molecule — the more highly screened the hydroxyl by substituents, the less the change in the spectrum. Data on the change in the spectrum on addition of alkali can be applied for the quantitative determination of these antioxidants by the differential method in presence of other substances which interfere in the direct determination [14]. For this reason, the spectra of antioxidants containing hydroxy groups and of phosphorous esters were determined both in neutral (solid line) and in alkaline (broken line) alcoholic media.

TABLE 4

SPECTRAL CHARACTERISTICS OF ANTIOXIDANTS – SECONDARY
AMINES

Figure	Antioxidant	Solvent	λ_{max}, mμ	a	ϵ
48	Neozone A (N-phenyl-1-naphthylamine)	Ethanol	253 340	79 39.5	17,300 8,650
49	Neozone D (N-phenyl-2-naphthylamine)	Ethanol	272.5 310 348	122 105 20	26,700 23,000 4,400
50	p-Hydroxyneozone [N-p-hydroxyphenyl-2-naphthylamine, p-(2-naphthylamino) phenol]	Ethanol	259 304	111 98	25,850 23,000
51	p-Hydroxydiphenylamine-(p-anilinophenol)	Ethanol	283	85	15,700
52	Agerite Stalite (octyl- or heptyl-substituted diphenylamine)	Ethanol	288	61	–
53	Antioxidant BLE (product of the condensation of diphenylamine with acetone)	Ethanol	289	95	–
54	Antioxidant BLE-25 (product of the condensation of diphenylamine with acetone)	Ethanol	287	105	–

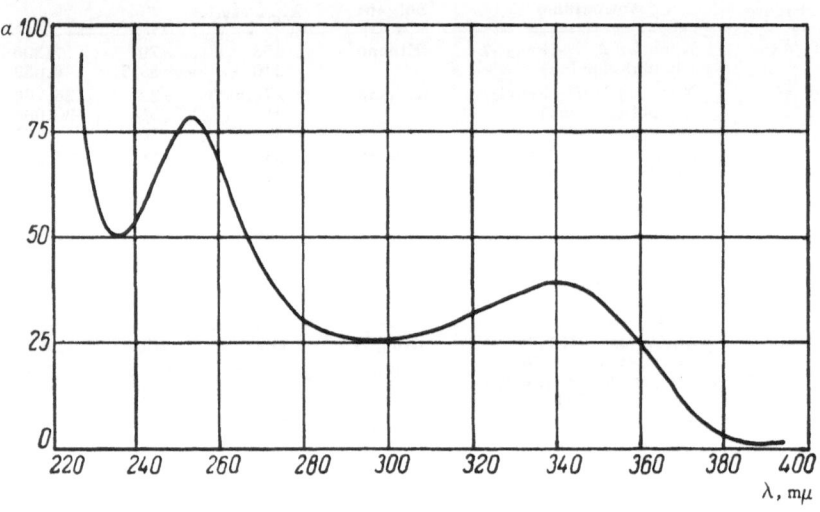

Fig. 48. Neozone A (*N*-phenyl-1-naphthylamine)

$C_{10}H_7NHC_6H_5$

Ethanol
C = 0.79 g/liter
d = 0.117 mm

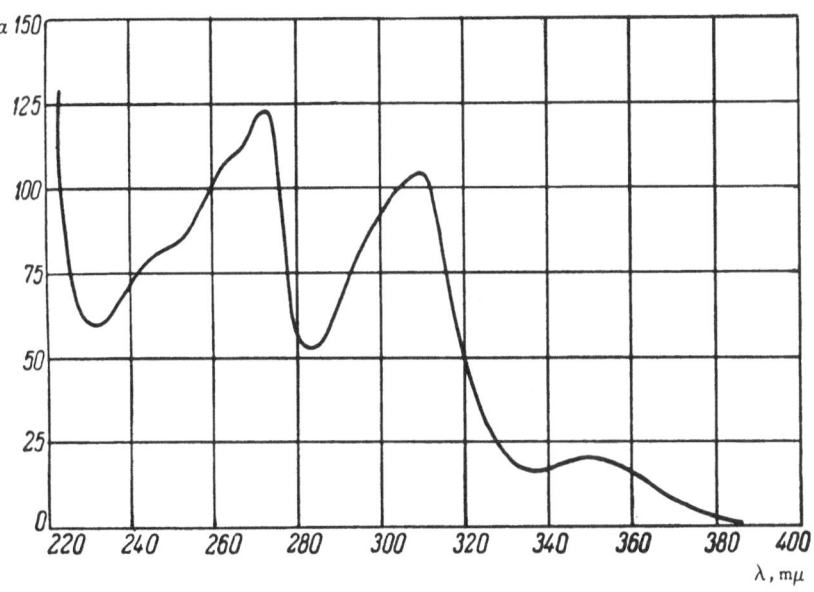

Fig. 49. Neozone D (*N*-phenyl-2-naphthylamine)

C₁₀H₇NHC₆H₅

Ethanol
C = 0.79 g/liter
d = 0.049 mm

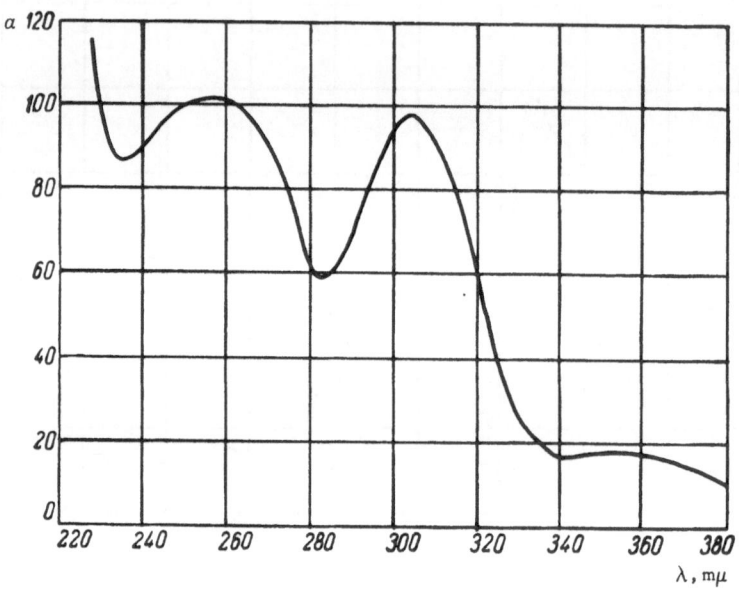

Fig. 50. p-Hydroxyneozone [N-p-hydroxyphenyl-2-naphthylamine, p-(2-naphthylamino) phenol]

$C_{10}H_7NHC_6H_4OH$

Ethanol
$C = 0.166$ g/liter
$d = 0.214$ mm

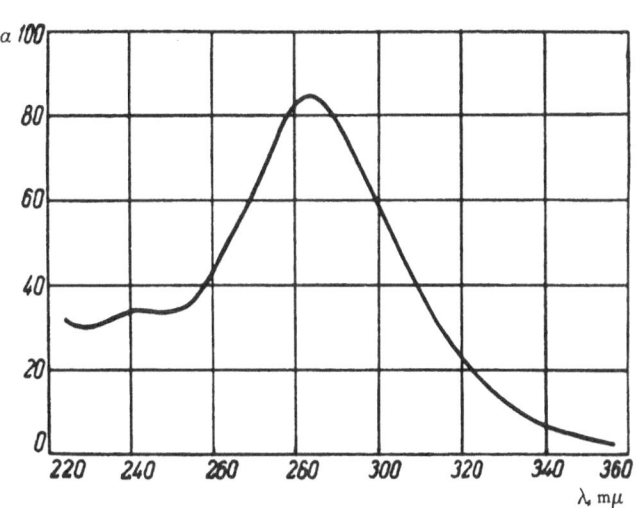

Fig. 51. *p*-Hydroxydiphenylamine (*p*-anilinophenol)

$C_6H_5 NHC_6H_4OH$

Ethanol
$C = 0.204$ g/liter
$d = 0.214$ mm

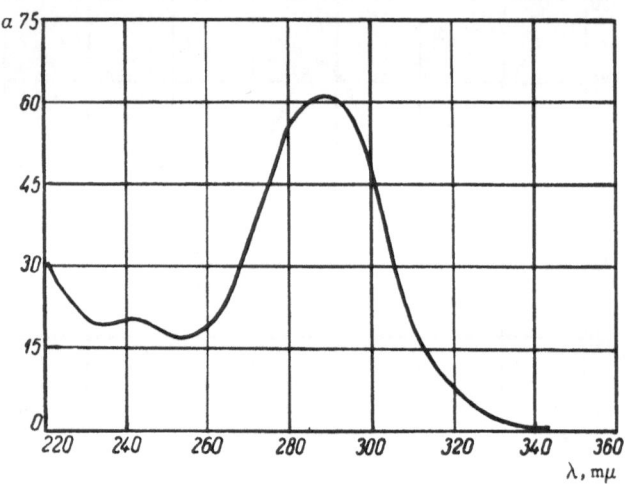

Fig. 52. Agerite Stalite

Ethanol
C = 2.60 g/liter
d = 0.66 mm

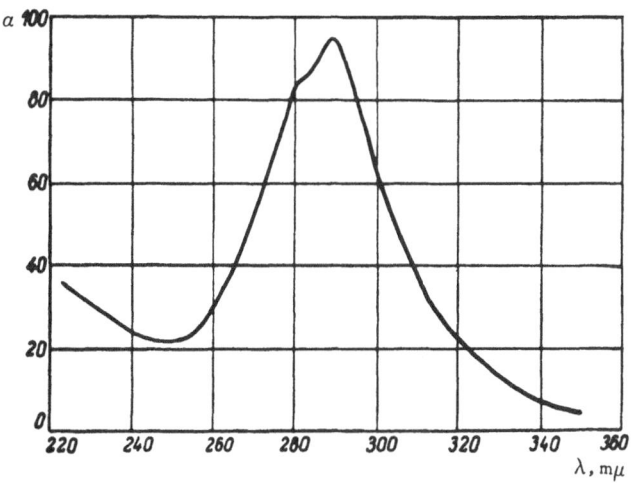

Fig. 53. Antioxidant BLE

Ethanol
$C = 1.00$ g/liter
$d = 0.205$ mm

65

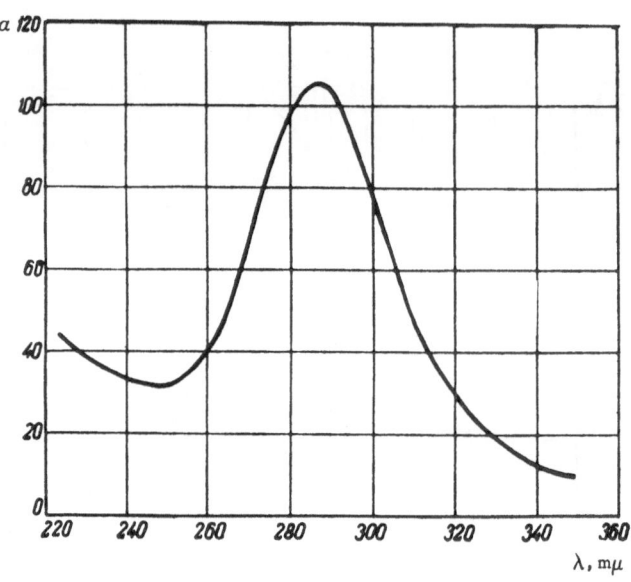

Fig. 54. Antioxidant BLE-25

Ethanol
$C = 1.00$ g/liter
$d = 0.107$ mm

TABLE 5
SPECTRAL CHARACTERISTICS OF ANTIOXIDANTS — p-ANISIDINE
DERIVATIVES

Figure	Antioxidant	Solvent	$\lambda_{max}, m\mu$	a	ϵ
55	N-Heptyl-p-anisidine	Ethanol	242	52	11,550
			308	12	2,650
56	N-Cyclohexyl-p-anisidine	Ethanol	242	73	15,000
			300	16.5	3,400
57	N-s-Octyl-p-anisidine	Ethanol	242	50	11,800
			302	10	2,350
58	N-α-Methylbenzyl-p-anisidine	Ethanol	244.5	52	15,100
			313	10	2,900
59	N-Isopropyl-p-anisidine	Ethanol	242	60	10,150
			300	11	1,850
60	N-Alkyl-p-anisidine	Ethanol	244	46	—
			310	8.5	

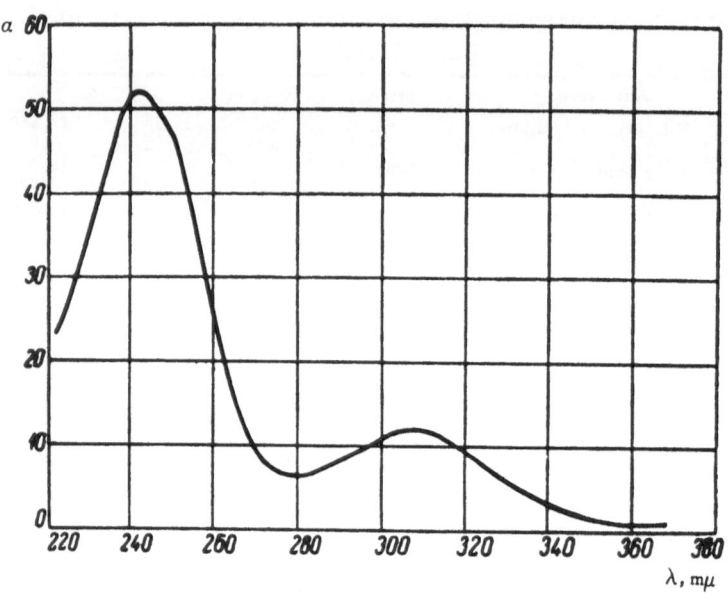

Fig. 55. *N*-Heptyl-*p*-anisidine

Ethanol
$C = 0.542$ g/liter
$d = 0.214$ mm $CH_3OC_6H_4NHC_7H_{15}$

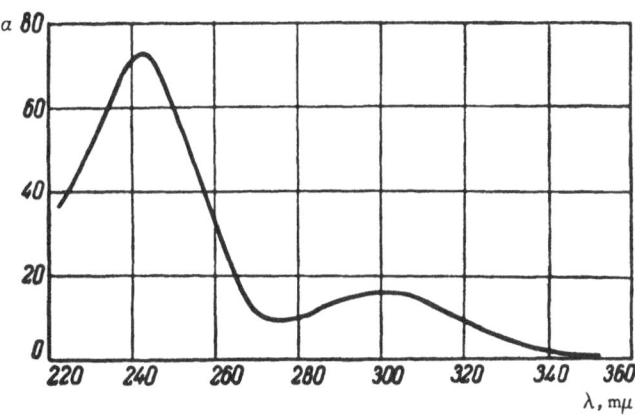

Fig. 56. *N*-Cyclohexyl-*p*-anisidine

$CH_3OC_6H_4NHC_6H_{11}$

Ethanol
$C = 0.592$ g/liter
$d = 0.214$ mm

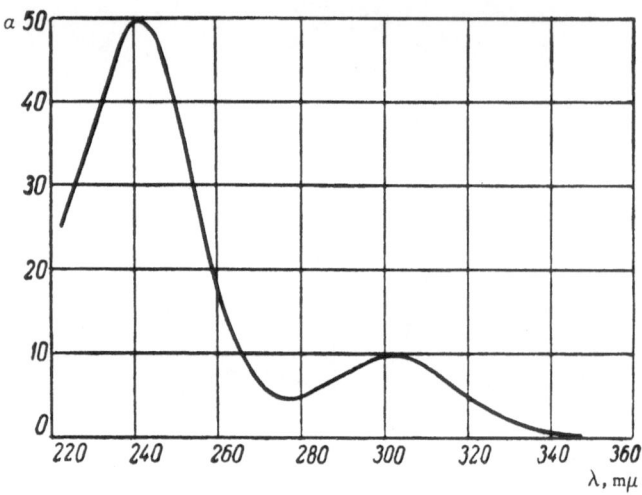

Fig. 57. *N*-s-Octyl-*p*-anisidine

$CH_3OC_6H_4NHC_8H_{17}$

Ethanol
$C = 0.650$ g/liter
$d = 0.214$ mm

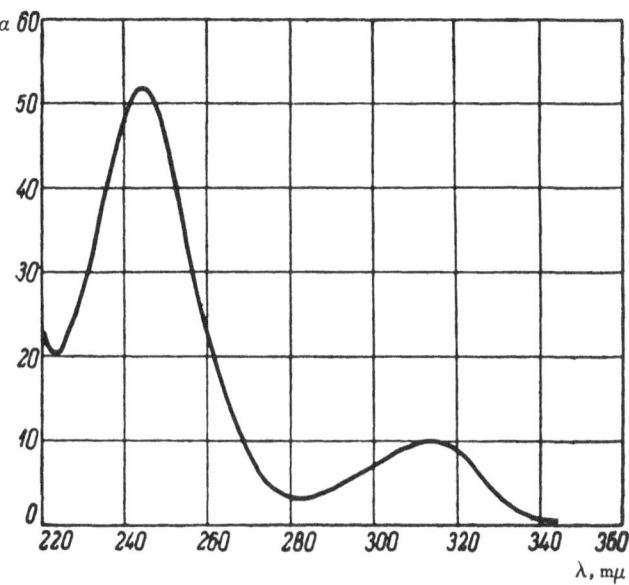

Fig. 58. *N-α*-Methylbenzyl-*p*-anisidine

$$\begin{array}{c}
C_6H_5 \\
| \\
CH_3OC_6H_4NHCH \\
| \\
CH_3
\end{array}$$

Ethanol
$C = 0.592$ g/liter
$d = 0.214$ mm

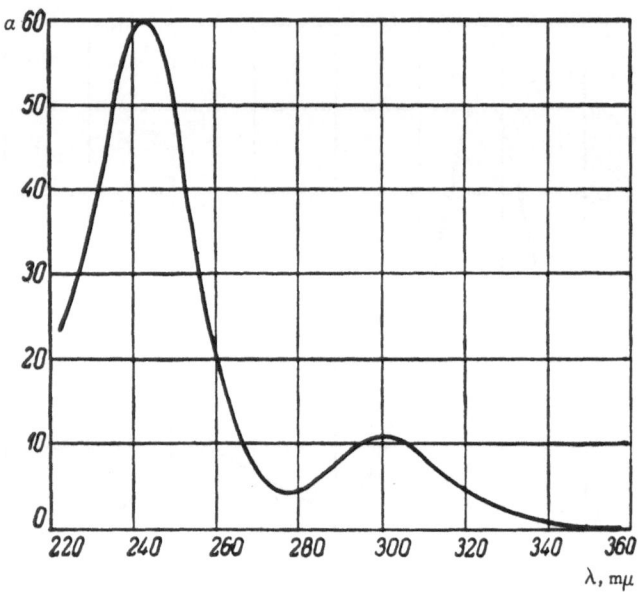

Fig. 59. *N*-Isopropyl-*p*-anisidine

$$CH_3$$
$$CH_3OC_6H_4NHCH$$
$$CH_3$$

Ethanol
$C = 0.536$ g/liter
$d = 0.214$ mm

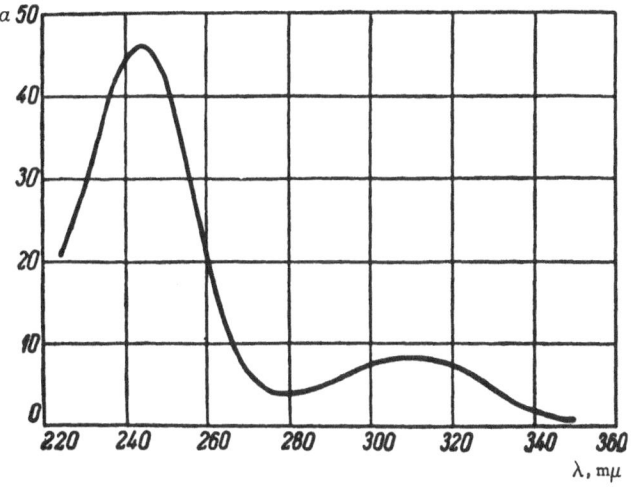

Fig. 60. *N*-Alkyl-*p*-anisidine

Ethanol
$C = 0.536$ g/liter
$d = 0.214$ mm

$CH_3OC_6H_4NHC_nH_{2n+1}$, $n = 7,8,9$

TABLE 6

SPECTRAL CHARACTERISTICS OF ANTIOXIDANTS – p-
PHENYLENEDIAMINE DERIVATIVES

Figure	Antioxidant	Solvent	λ_{max}, mμ	α	ϵ
61	Antioxidant DFFD(N,N'-diphenyl-p-phenylenediamine)	Ethanol	304	105	27,300
62	Antioxidant 4010 (N-cyclohexyl-N'-phenyl-p-phenylenediamine)	Ethanol Ethanol	290	70	18,600
63	Antioxidant 4010-NA (N-isopropyl-N'-phenyl-p-phenylenediamine)	Ethanol	290	83	18,850
64	N-Pentyl-N'-phenyl-p-phenylenediamine	Ethanol	290	53.5	13,650
65	N-Hexyl-N'-phenyl-p-phenylenediamine	Ethanol	290	77	20,700
66	N-Octyl-N'-phenyl-p-phenylenediamine	Ethanol	290	55	16,350
67	N-Decyl-N'-phenyl-p-phenylenediamine	Ethanol	290	54	17,550
68	N-Alkyl-N'-phenyl-p-phenylenediamine (technical product)	Ethanol	290	55	—
69	N,N'-Dipentyl-p-phenylenediamine	Ethanol	259	27.5	6,850
70	N,N'-Dioctyl-p-phenylenediamine	Ethanol	259	38.5	12,800
71	N,N'-Dinonyl-p-phenylenediamine	Ethanol	259	33	11,900
72	Agerite white (N,N'-di-2-naphthyl-p-phenylenediamine)	Ethanol 80% + chloroform (20%)	246 274 318	74 63 69.5	27,000 23,000 25,300
73	Aranox [N-phenyl-N'-(p-tolylsulfonyl)-p-phenylenediamine,N-(p-anilinophenyl)-p-toluenesulfonamide]	Ethanol	227 293	46 54	15,550 18,250

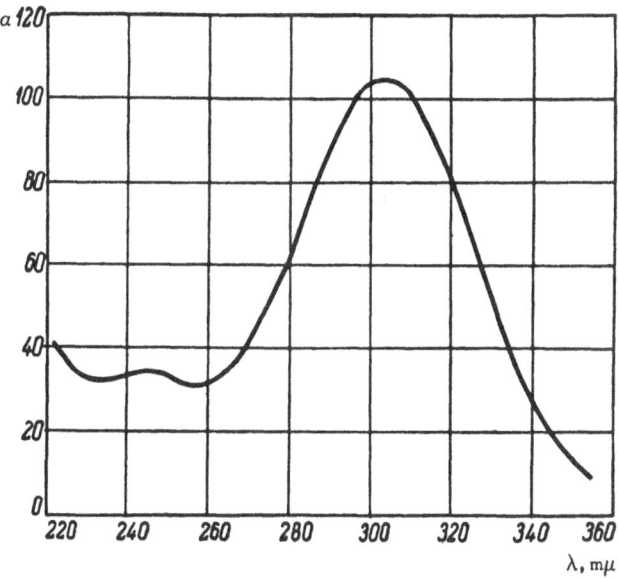

Fig. 61. Antioxidant DFFD (*N,N'*-diphenyl-*p*-phenylenediamine)

$C_6H_5\,NHC_6H_4NHC_6H_5$

Ethanol
$C = 0.10$ g/liter
$d = 0.107$ mm

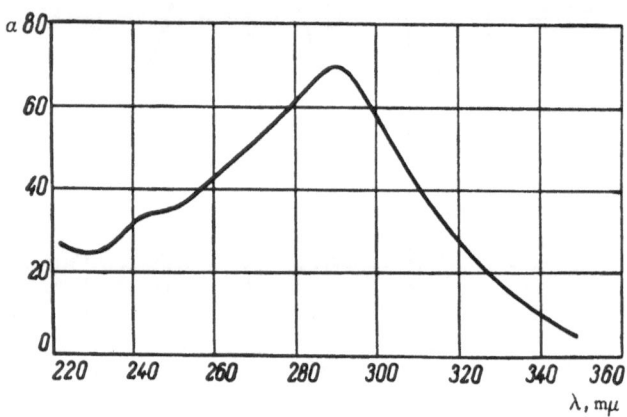

Fig. 62. Antioxidant 4010 (*N*-cyclohexyl-*N'*-phenyl-*p*-phenylenediamine)

C₆H₅ NHC₆H₄NHC₆H₁₁

Ethanol
$C = 2.00$ g/liter
$d = 0.066$ mm

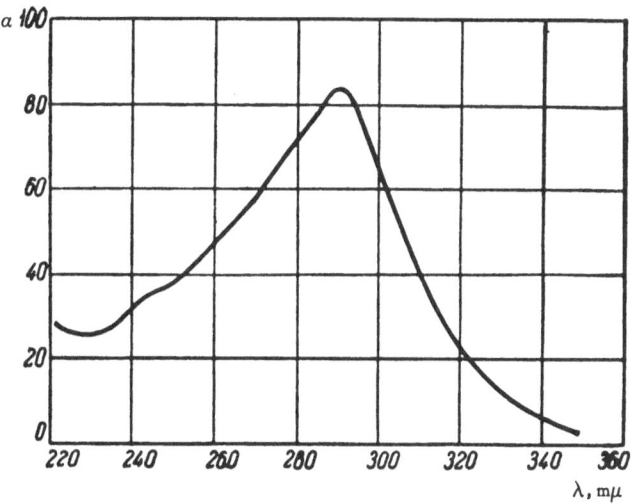

Fig. 63. Antioxidant 4010-NA (*N*-isopropyl-*N'*-phenyl-*p*-phenylenediamine)

$$C_6H_5\,NHC_6H_4NHCH \begin{array}{c} CH_3 \\ | \\ \\ | \\ CH_3 \end{array}$$

Ethanol
$C = 1.00$ g/liter
$d = 0.066$ mm

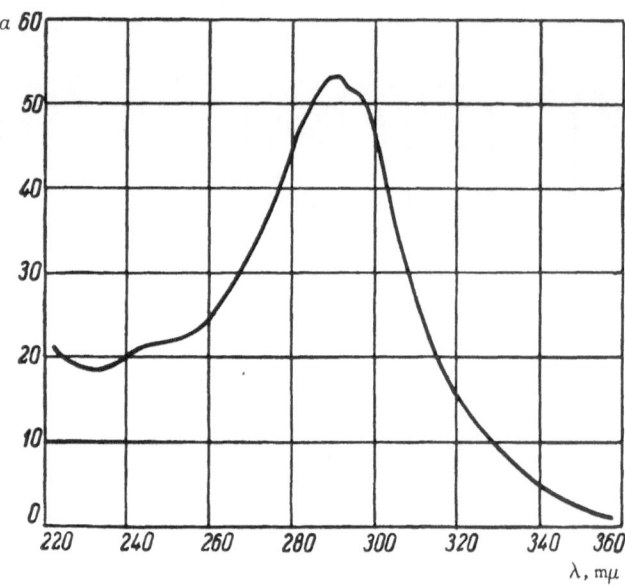

Fig. 64. *N*-Pentyl-*N'*-phenyl-*p*-phenylenediamine

$C_6H_5NHC_6H_4NHC_5H_{11}$

Ethanol
$C = 0.410$ g/liter
$d = 0.214$ mm

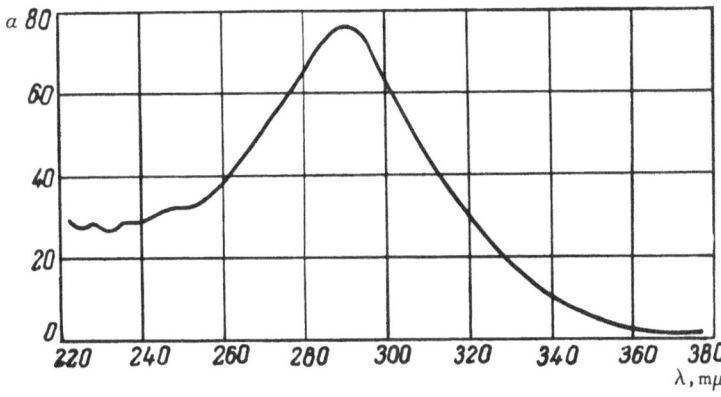

Fig. 65. *N*-Hexyl-*N'*-phenyl-*p*-phenylenediamine

$C_6H_5NHC_6H_4NHC_6H_{13}$

Ethanol
$C = 0.472$ g/liter
$d = 0.117$ mm

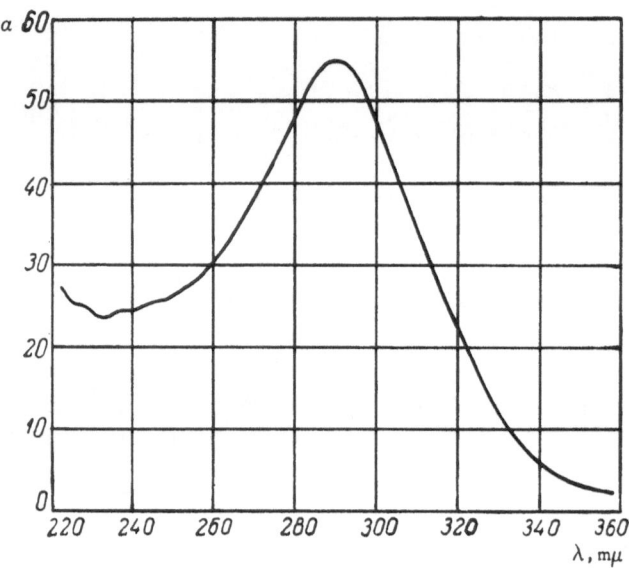

Fig. 66. *N*-Octyl-*N'*-phenyl-*p*-phenylenediamine

$C_6H_5\,NHC_6H_4NHC_8H_{17}$

Ethanol
$C = 0.460$ g/liter
$d = 0.117$ mm

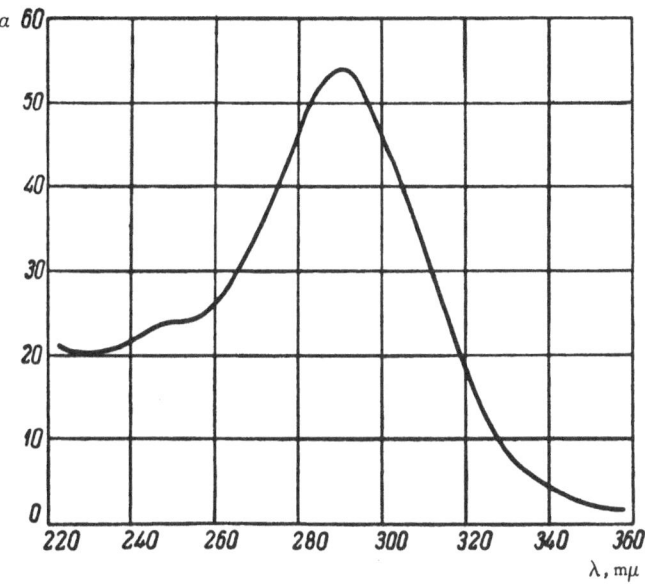

Fig. 67. *N*-Decyl-*N'*-phenyl-*p*-phenylenediamine

$C_6H_5NHC_6H_4NHC_{10}H_{21}$

Ethanol
$C = 0.240$ g/liter
$d = 0.509$ mm

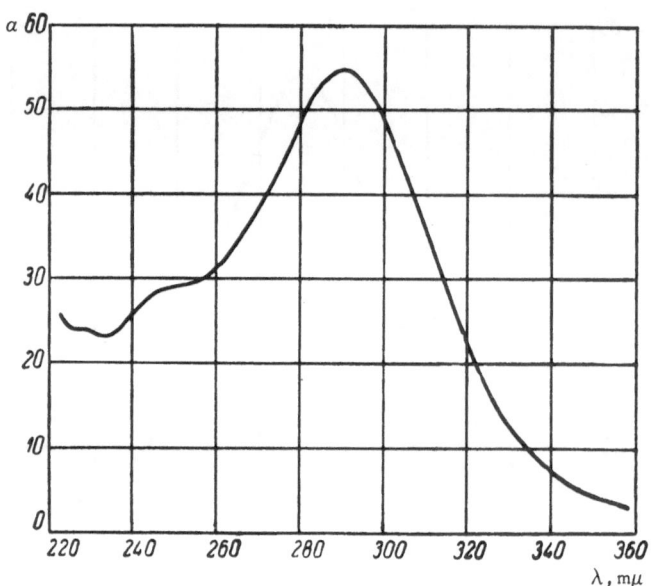

Fig. 68. N-Alkyl-N'-phenyl-p-phenylenediamine (technical product)

Ethanol
C = 0.636 g/liter
$C_6H_5\,NHC_6H_4NHC_nH_{2n+1}$, $n = 7,8,9$ d = 0.117 mm

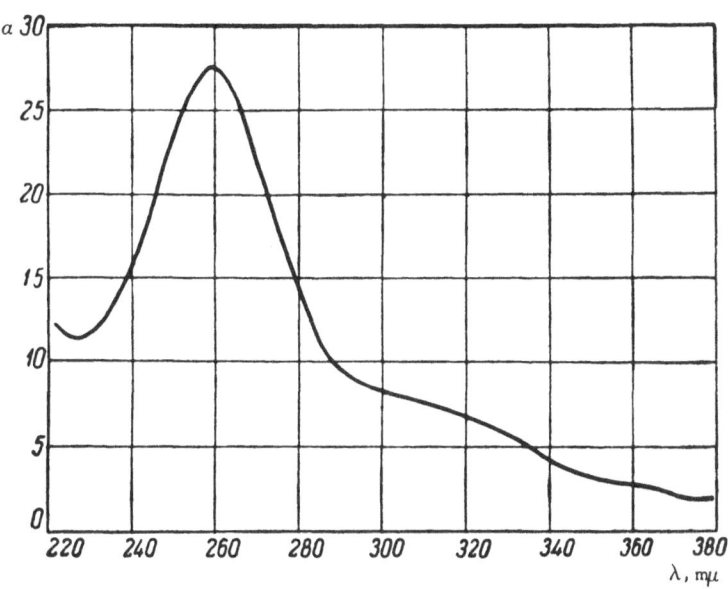

Fig. 69. N,N'-Dipentyl-p-phenylenediamine

$C_5 H_{11} NHC_6 H_4 NHC_5 H_{11}$

Ethanol
$C = 0.3376$ g/liter
$d = 0.509$ mm

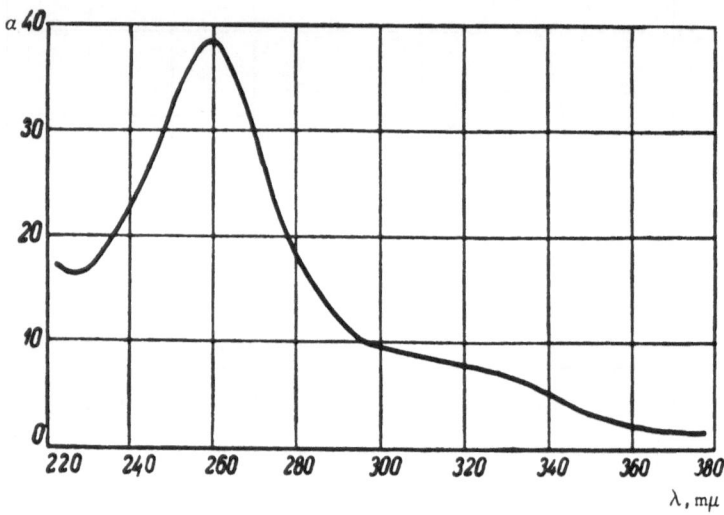

Fig. 70. N,N'-Dioctyl-p-phenylenediamine

$C_8H_{17}NHC_6H_4NHC_8H_{17}$

Ethanol
$C = 0.492$ g/liter
$d = 0.509$ mm

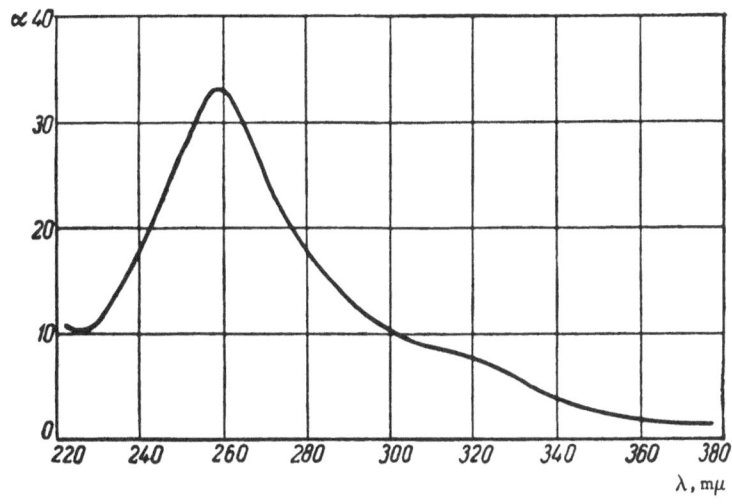

Fig. 71. N,N'-Dinonyl-p-phenylenediamine

$C_9H_{19}NHC_6H_4NHC_9H_{19}$

Ethanol
$C = 0.398$ g/liter
$d = 0.214$ mm

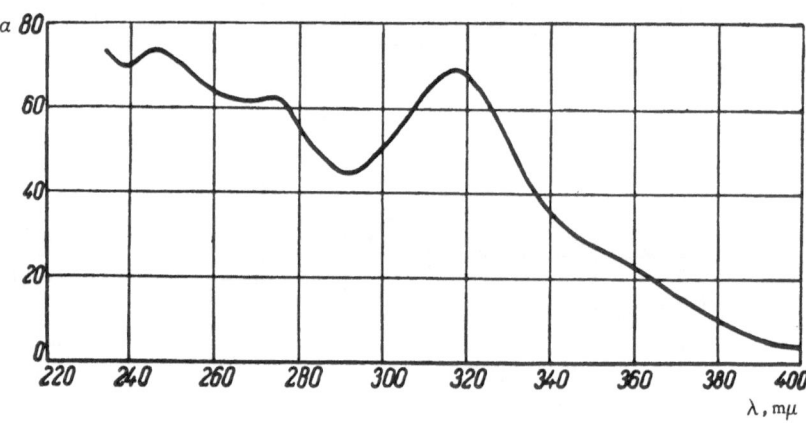

Fig. 72. Agerite white (*N*,*N'*-di-2- naphthyl-*p*-phenylenediamine)

C₁₀H₇NHC₆H₄NHC₁₀H₇

80% ethanol, 20% chloroform
$C = 0.1244$ g/liter
$d = 0.509$ mm

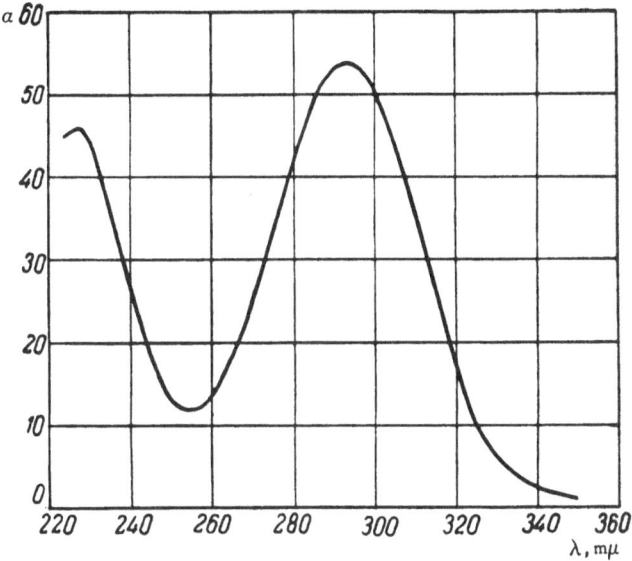

Fig. 73. Aranox [N-phenyl-N'-(p-tolylsulfonyl)-p-phenylenediamine, N-(p-anilinophenyl)-p-toluenesulfonamide]

CH₃C₆H₄SO₂NHC₆H₄NHC₆H₅

Ethanol
C = 0.0231 g/liter
d = 5.01 mm

TABLE 7

SPECTRAL CHARACTERISTICS OF ANTIOXIDANTS – PHENOL
DERIVATIVES

Figure	Antioxidant	Neutral alcoholic solution			Solution in 0.1 N alcoholic alkali		
		λ_{max}, mμ	α	ϵ	λ_{max}, mμ	α	ϵ
74	Ionol, Antioxidant P-21 (2,6-di-t-butyl-p-cresol)	278	8.0	1,760	278 307	8.0 2.0	1,760 440
75	Antioxidant P-23 (2,4,6-tri-t-butylphenol)	274.5	7.0	1,760	274.5 302	7.0 2.0	1,760 500
76	Antioxidant 425 (2,2'-methylenebis-[6-t-butyl-4-ethylphenol])	282	13.0	4,780	286.5 307	13.0 17.0	4,780 6,250
77	Antioxidant 2246 (2,2'-methylenebis-[6-t-butyl-p-cresol])	281	12.0	4,080	287 308	12.0 17.0	4,080 5,780
78	Bukremet (4,4'-methylenebis [6-t-butyl-o-cresol])	279	10.7	3,600	255 293	31.7 14.0	10,700 4,750
79	Santowhite powder (4,4'-butylidenebis-[6-t-butyl-m-cresol])	281	1.3	5,000	287	13.5	5,150
80	p,p'-Dihydroxybiphenyl (p,p'-biphenol)	264	126	23,500	293	153	28,500
81	Wing-Stay S ⎫ Mixtures of mono-, di-, and	280	8.5	–	302	15.0	–
82	Montaclere ⎬ tri-styryl derivatives of	279.5	8.0	–	303	18.2	–
83	Styphen I ⎭ phenols	282	6.2	–	306	10.2	–
84	Nonox EX ⎫ Products of condensation	281	14.0	–	302	19.5	–
85	Nonox EXN ⎬ of phenols with aldehydes	283	17.0	–	301	21.0	–
86	Nonox WSL ⎬	280	9.3	–	287 301	9.5 9.3	–
87	Nonox WSP ⎭	283	11.5	–	288 310	11.7 15.3	–
88	Agerite Superlite (polyalkylated poly-phenol)	280	11.2	–	285	11.0	–

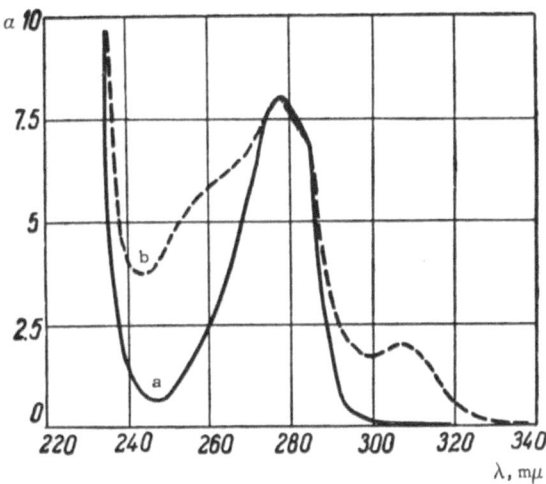

Fig. 74. Ionol, Antioxidant P-21 (2,6-di-t-butyl-p-cresol)

Ethanol (a), 0.1 N ethanolic KOH (b)
$C = 3.67$ g/liter
$d = 0.214$ mm

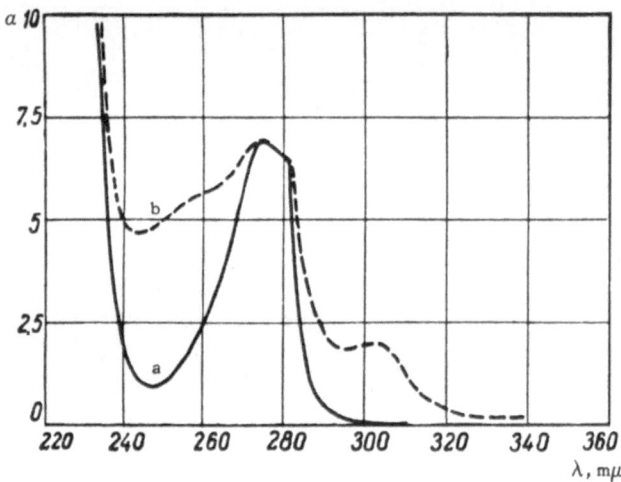

Fig. 75. Antioxidant P-23 (2,4,6-tri-t-butylphenol)

Ethanol (a), 0.1 N ethanolic KOH (b)
$C = 14.00$ g/liter
$d = 0.049$ mm

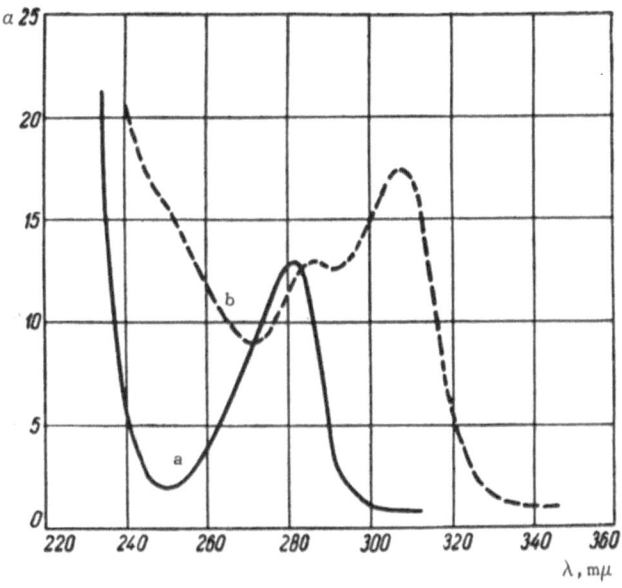

Fig. 76. Antioxidant 425 (2,2'-methylenebis[6-t-butyl-4-ethylphenol])

Ethanol(a), 0.1 N ethanolic KOH (b)
C = 7.90 g/liter
d = 0.066 mm

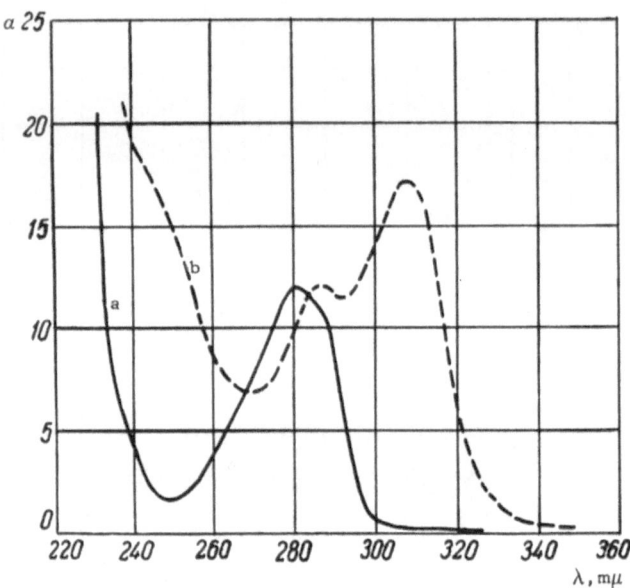

Fig. 77. Antioxidant 2246 (2,2′-methylenebis[6-t-butyl-p- cresol])

$(CH_3)_3 C$ — OH — CH_2 — OH — $C(CH_3)_3$
     CH_3 CH_3

Ethanol (a), 0.1 N ethanolic KOH (b)
$C - 1.40$ g/liter
$d = 0.214$ mm

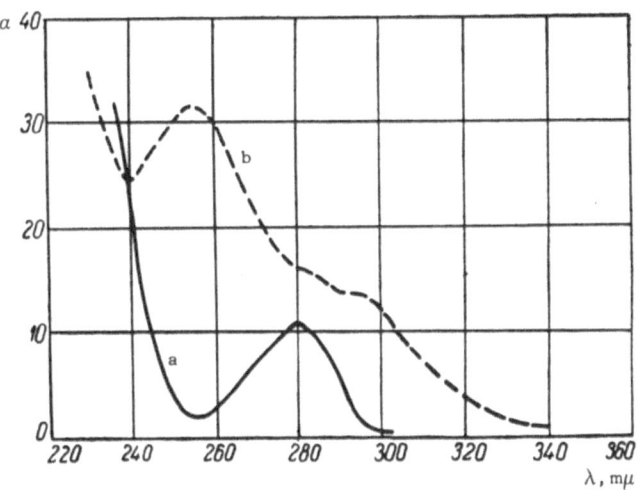

Fig. 78. Bukremet (4,4'-methylenebis [6-t-butyl-o-cresol])

Ethanol (a), 0.1 N ethanolic KOH (b)
$C = 0.734$ g/liter
$d = 0.509$ mm

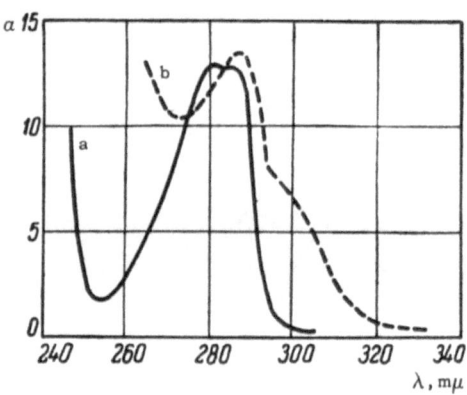

Fig. 79. Santowhite powder (4,4′-butylidenebis [6-t-butyl-*m*-cresol])

HO—⟨ring⟩—CH—⟨ring⟩—OH
  C(CH₃)₃   C(CH₃)₃
       C₃H₇
   CH₃   CH₃

Ethanol (a), 0.1 N ethanolic KOH (b)
C = 0.800 g/liter
d = 0.509 mm

94

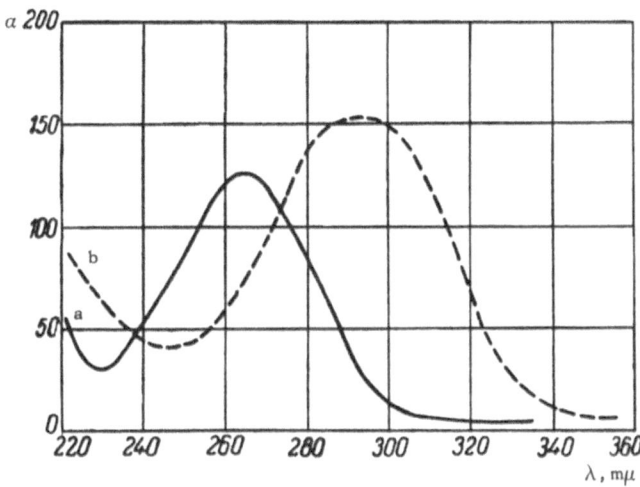

Fig. 80. *p,p'*-Dihydroxybiphenyl (*p,p'*-biphenol)

HOC$_6$H$_4$C$_6$H$_4$OH

Ethanol (a), 0.1 N ethanolic KOH(b)
C = 0.79 g/liter
d = 0.066 mm

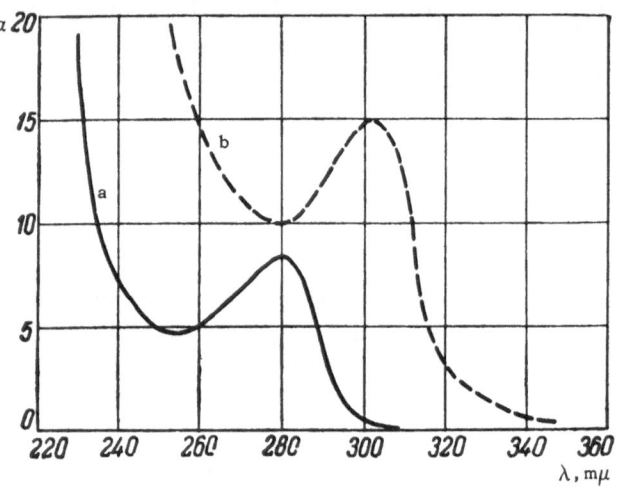

Fig. 81. Wing-Stay S

Ethanol (a) 0.1 N ethanolic KOH (b)
$C = 0.78$ g/liter
$d = 0.214$ mm

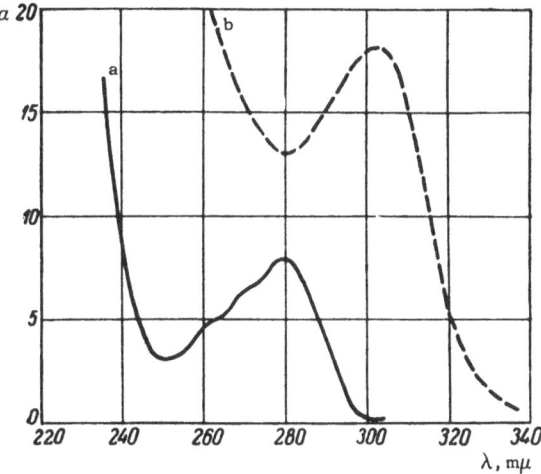

Fig. 82. Montaclere

Ethanol (a), 0.1 N ethanolic KOH (b)
$C = 0.82$ g/liter
$d = 0.214$ mm

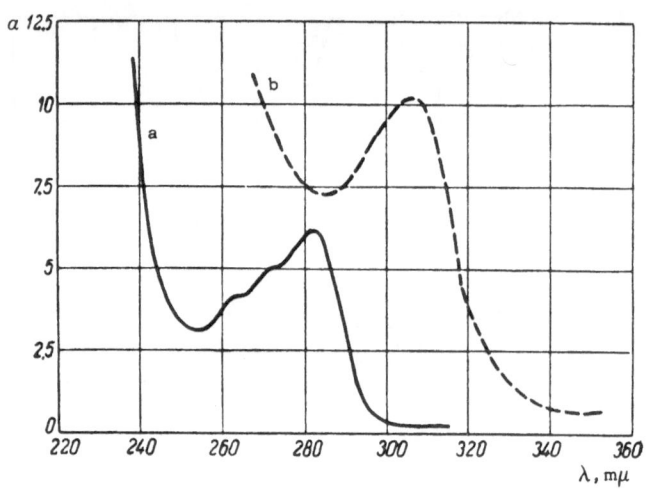

Fig. 83. Styphen I

Ethanol (a), 0.1 N ethanolic KOH (b)
C = 10.0 g/liter
d = 0.066 mm

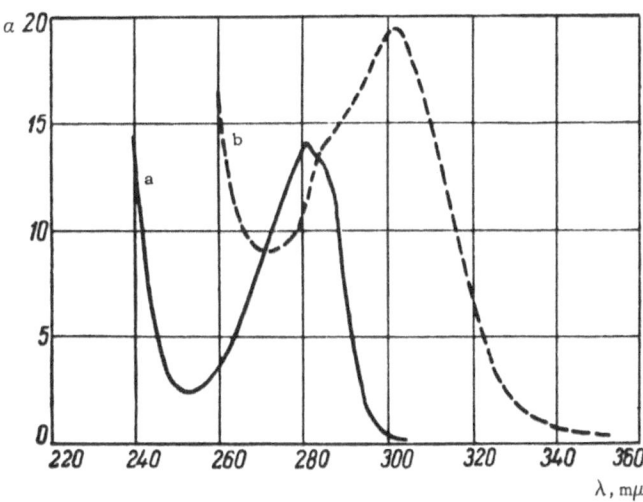

Fig. 84. Nonox EX

Ethanol (a), 0.1 N ethanolic KOH (b)
$C = 0.1006$ g/liter
$d = 5.01$ mm

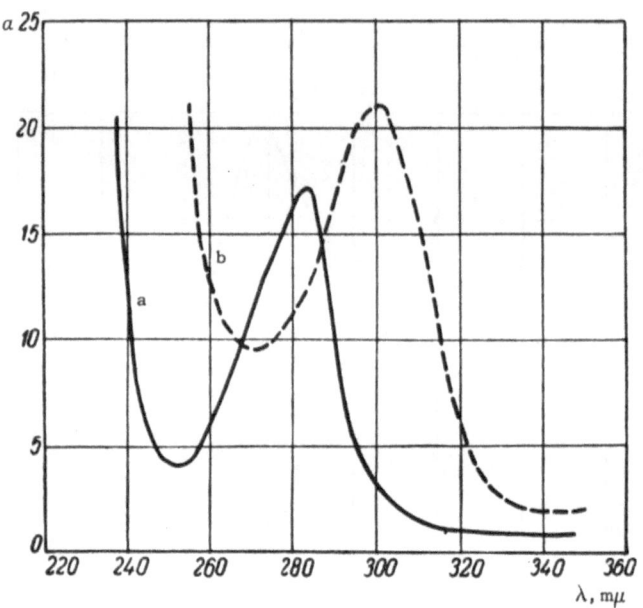

Fig. 85. Nonox EXN

Ethanol (a), 0.1 N ethanolic KOH (b)
$C = 7.9$ g/liter
$d = 0.066$ mm

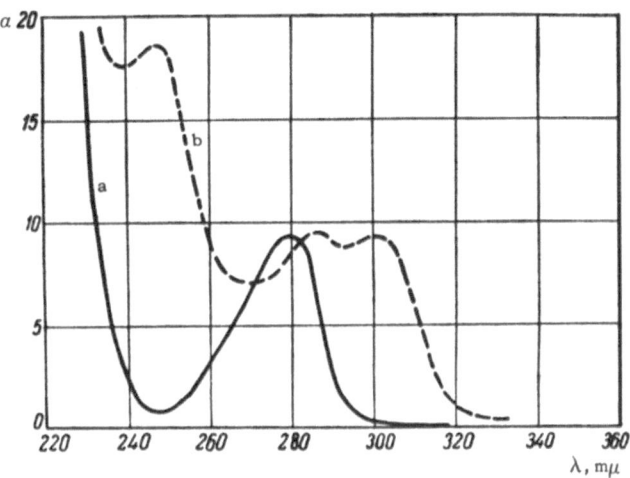

Fig. 86. Nonox WSL

Ethanol (a), 0.1 N ethanolic KOH (b)
C = 10 g/liter
d = 0.066 mm

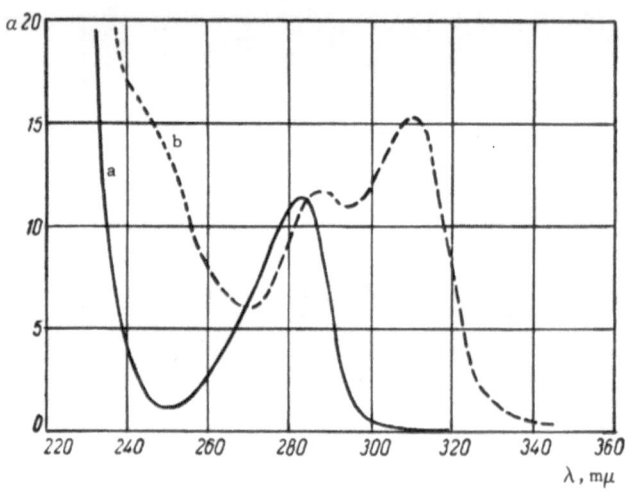

Fig. 87. Nonox WSP

Ethanol (a), 0.1 N ethanolic KOH (b)
$C = 10$ g/liter
$d = 0.066$ mm

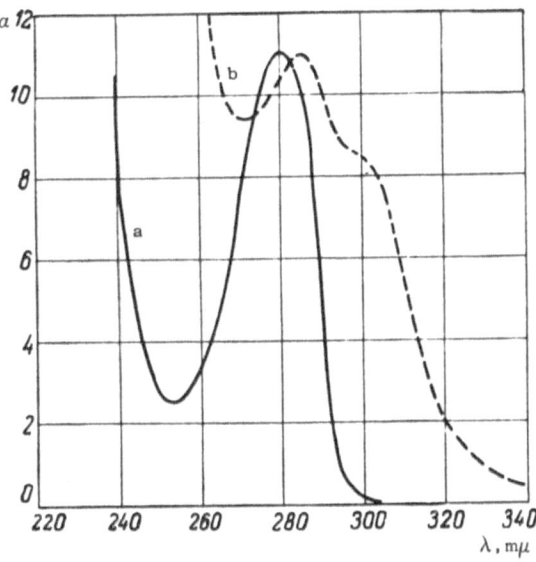

Fig. 88. Agerite Superlite (polyalkylated polyphenol)

Ethanol (a), 0.1 N ethanolic KOH (b)
$C = 0.722$ g/liter
$d = 0.509$ mm

TABLE 8

SPECTRAL CHARACTERISTICS OF ANTIOXIDANTS – PHOSPHOROUS
ESTERS

Figure	Antioxidant	Neutral alcoholic solution			Solution in 0.1 N alcoholic alkali		
		λ_{max}, mμ	α	ϵ	λ_{max}, mμ	α	ϵ
89	Polygard (tris-p-nonylphenyl phosphite)	272	4.0	2,750	295	8.5	5,850
90	Ionol pyrocatechol phosphite (2,6-di-t-butyl-p-tolyl-o-phenylene phosphite)	272 277	11.0 9.5	3,950 2,500	Destroyed		
91	Phosphite, Antioxidant P-24 (phosphorous ester of styryl-substituted phenol)	269 279	6.2 6.8	– –	300	12.2	–

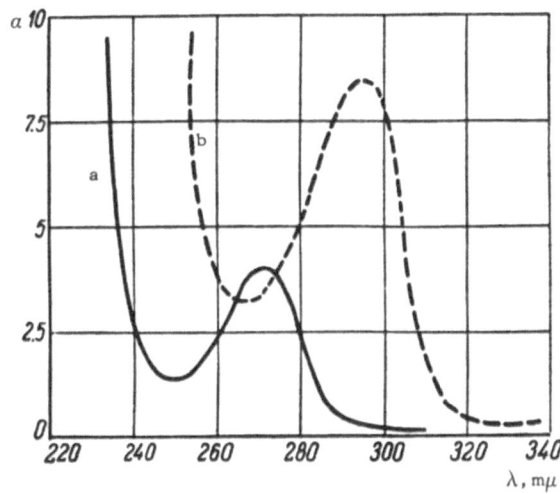

Fig. 89. Polygard (tris-*p*-nonylphenyl phosphite)

$(C_9H_{19}C_6H_4O-)_3P$

Ethanol (a), 0.1 N ethanolic KOH (b)
$C = 4.0$ g/liter
$d = 0.214$ mm

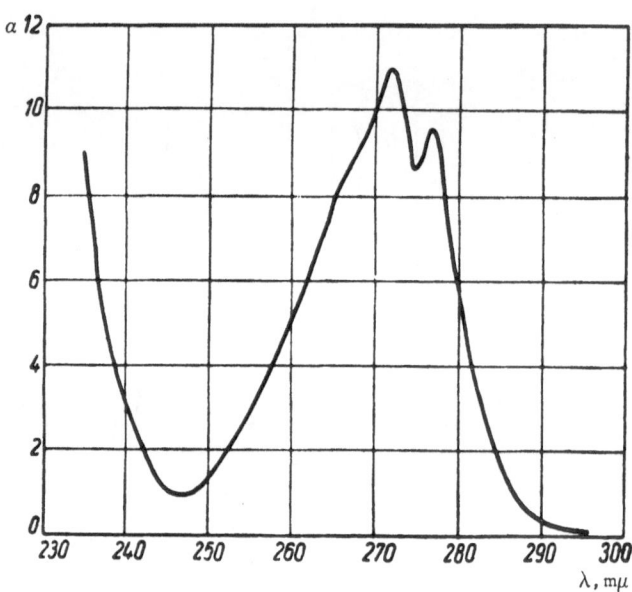

Fig. 90. Ionol pyrocatechol phosphite (2,6-di-t-butyl-*p*-tolyl-*o*-phenylene phosphite)

Ethanol
$C = 0.1204$ g/liter
$d = 5.01$ mm

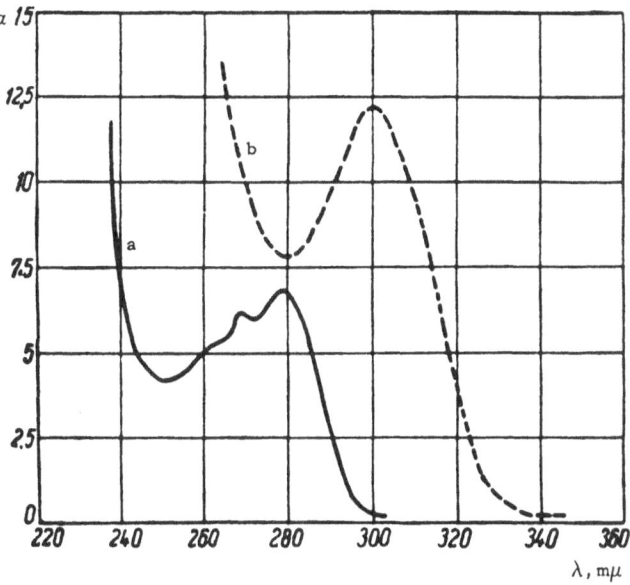

Fig. 91. Phosphite, Antioxidant P-24

Ethanol (a), 0.1 N ethanolic KOH (b)
C = 0.164 g/liter
d = 5.01 mm

TABLE 9

SPECTRAL CHARACTERISTICS OF ANTIOXIDANTS – HYDROQUINONE
DERIVATIVES

Figure	Antioxidant	Neutral alcoholic solution			Solution in 0.1 N alcoholic alkali		
		λ_{max}, mμ	α	ϵ	λ_{max}, mμ	α	ϵ
92	Agerite Alba [p-(benzyloxy)-phenol]	291	13	2,600	308	14	2,800
93	Santovar O, Antioxidant P-20 (2,5-di-t-butylhydroquinone)	294	21	4,650	252	27	6,000
					324	47	10,400
94	Santovar A (2,5-di-t-pentylhydroquinone)	230	18	4,500	256	22	5,500
		294	19	4,750	324.5	35	8,750

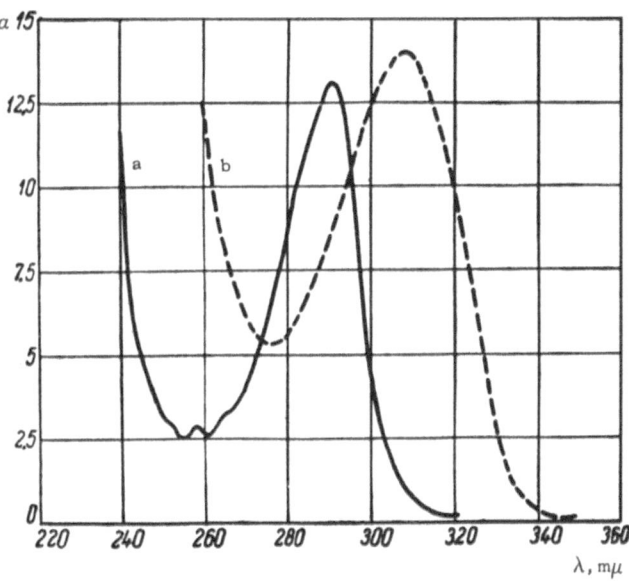

Fig. 92. Agerite Alba [p-(benzyloxy)phenol]

C₆H₅ CH₂OC₆H₄OH

Ethanol (a), 0.1 N ethanolic KOH (b)
$C = 0.682$ g/liter
$d = 0.509$ mm

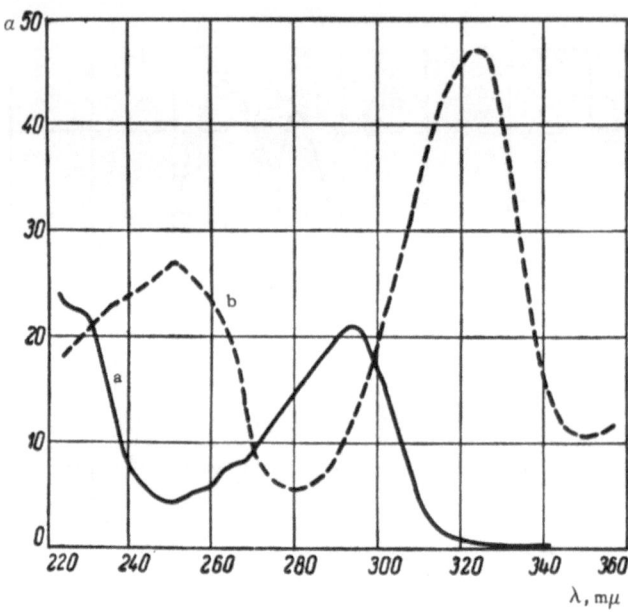

Fig. 93. Santovar O, Antioxidant P-20 (2,5-di-t-butylhydroquinone)

Ethanol (a), 0.1 N ethanolic KOH (b)
$C = 7.9$ g/liter
$d = 0.066$ mm

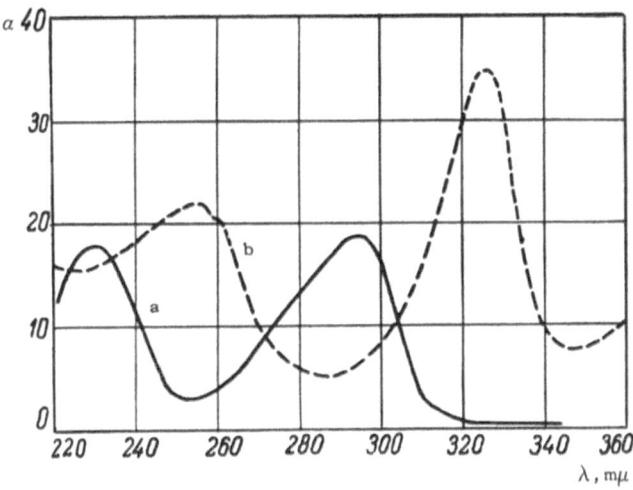

Fig. 94. Santovar A (2,5-di-t-pentylhydroquinone)

Ethanol (a), 0.1 N ethanolic KOH (b)
C = 7.9 g/liter
d = 0.066 mm

TABLE 10

SPECTRAL CHARACTERISTICS OF ANTIOXIDANTS — SULFIDES

Figure	Antioxidant	Neutral alcoholic solution			Solution in 0.1 N alcoholic alkali		
		λ_{max}, mμ	α	ϵ	λ_{max}, mμ	α	ϵ
95	Antioxidant SAO-6 (2,2'-thiobis[6-t-butyl-p-cresol])	292	14.5	5,200	298 319	17.2 18.5	6,150 6,600
96	Antioxidant AN-6 (2,2' thiobis [6-(α,α-dimethyl-benzyl)-p-cresol])	294	7.0	3,375	298 317	5.5 6.5	2,650 3,150
97	Santowhite crystals (4,4'-thiobis [6-t-butyl-m-cresol])	249 281	47.0 21.5	16,800 7,700	269	59.0	21,100
98	Santowhite CM (6,6'-thiodi-o-cresol)	236 251	56.0 60.0	13,800 14,750	266	108.0	26,600
99	Santowhite MK (product of reaction of sulfur chloride with 6-t-butyl-m-cresol)	284	19.0	—	251	37.5	—

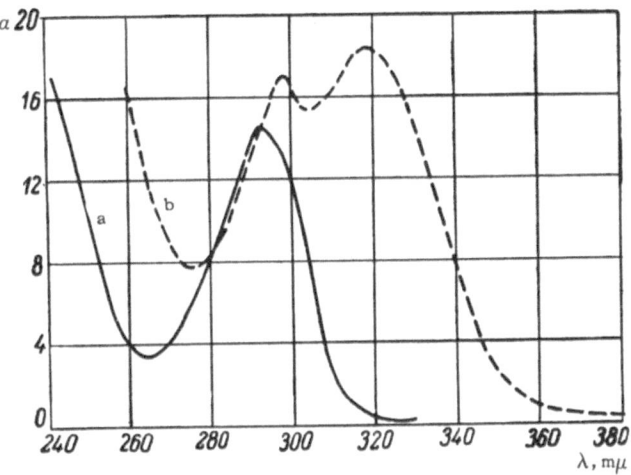

Fig. 95. Antioxidant SAO-6 (2,2'-thiobis [6-t-butyl-p-cresol])

Ethanol (a), 0.1 N ethanolic KOH (b)
$C = 0.6112$ g/liter
$d = 0.509$ mm

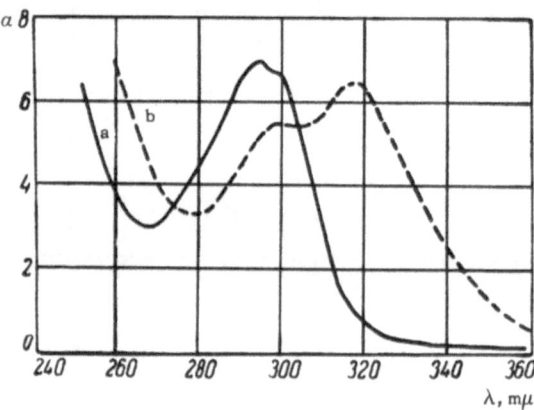

Fig. 96. Antioxidant AN-6 (2,2'-thiobis[6-(α,α-dimethylbenzyl)-p-cresol])

$C_6H_5(CH_3)_4C$ — OH ... S ... OH — $C(CH_3)_3C_6H_5$

CH₃   CH₃

Ethanol (a), 0.1 N ethanolic KOH (b)
$C = 0.777$ g/liter
$d = 1.016$ mm

114

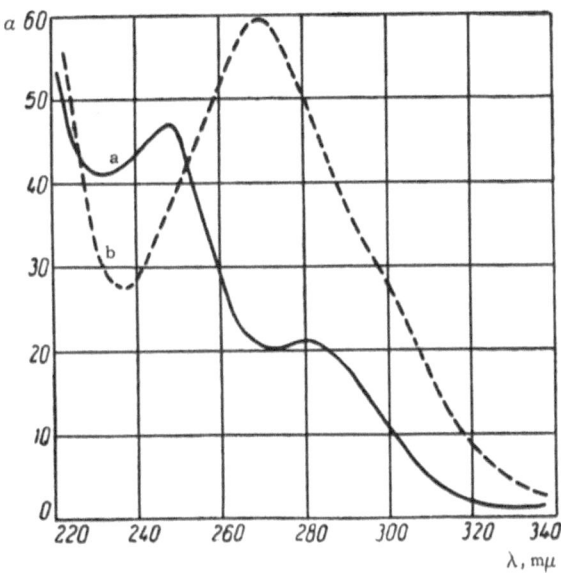

Fig. 97. Santowhite crystals (4,4′-thiobis[6-t-butyl-m-cresol])

Ethanol (a), 0.1 N ethanolic KOH (b)
C = 0.1004 g/liter
d = 1.016 mm

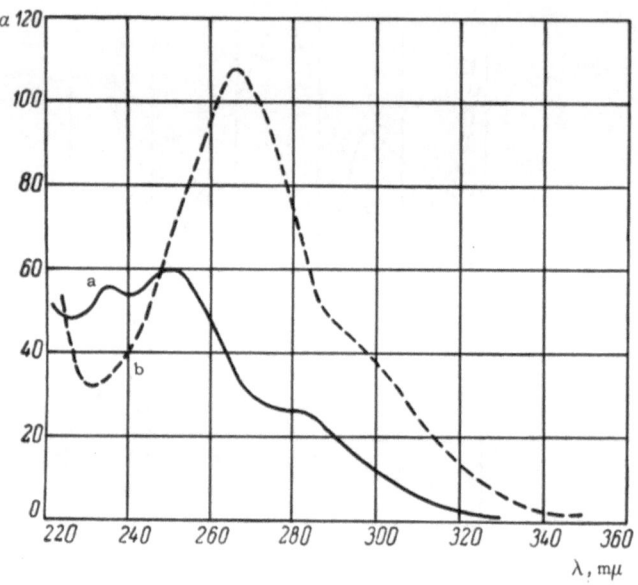

Fig. 98. Santowhite CM (6,6'-thiodi-*o*-cresol)

Ethanol (a), 0.1 N ethanolic KOH (b)
$C = 0.112$ g/liter
$d = 1.016$ mm

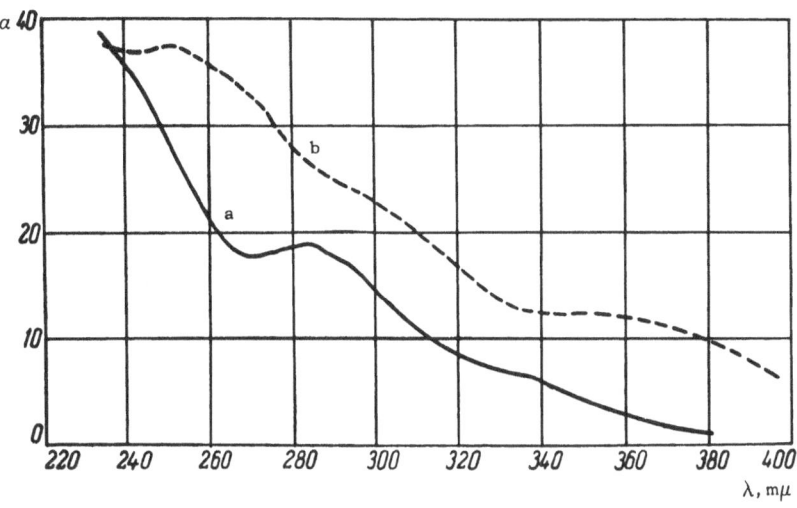

Fig. 99. Santowhite MK

Ethanol (a), 0.1 N ethanolic KOH (b)
C = 0.796 g/liter
d = 0.214 mm

TABLE 11

SPECTRAL CHARACTERISTICS OF ANTIOXIDANTS – HYDROQUINOLINE
DERIVATIVES

Figure	Antioxidant	Solvent	λ_{max}, mμ	α	ϵ
100	Santoflex AW (6-ethoxy-1,2-dihydro-2,2,4-trimethylquinoline)	Ethanol	229 350	100 11	21,800 2,400
101	Santoflex DD (6-dodecyl-1,2-dihydro-2,2,4-trimethylquinoline)	Ethanol	233 342	78 5	26,600 1,700
102	Agerite Resin D	Ethanol	236 296	89 13	15,400 2,250
103	Flectol B	Ethanol	232 310 322	143 12 12	24,750 2,075 2,075
104	Antox (product of condensation of aniline with butyraldehyde)	Ethanol	231 235 248 305 318	98 102 32 15 15	– – – – –
105	Antioxidant I-18 (product of condensation of aniline with isoprene)	Ethanol	247 299	60 15	– –

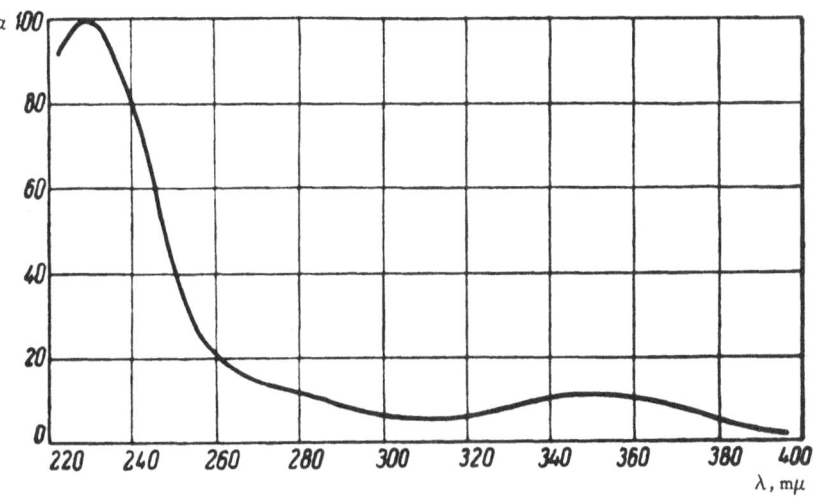

Fig. 100. Santoflex AW (6-ethoxy-1,2-dihydro-2,2,4-trimethylquinoline)

Ethanol
$C = 0.1572$ g/liter
$d = 0.509$ mm

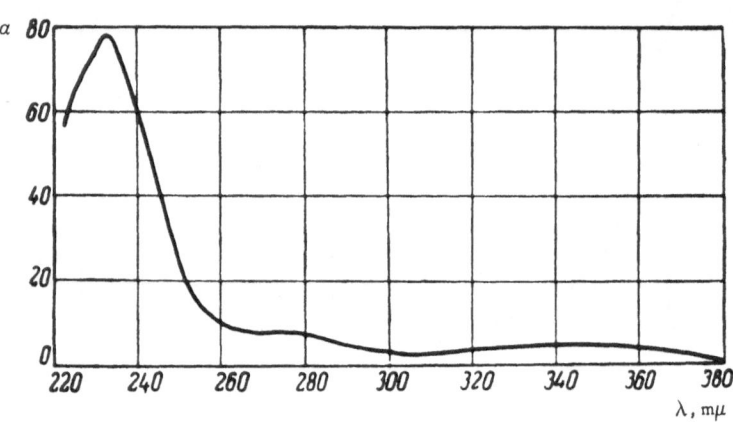

Fig. 101. Santoflex DD (6-dodecyl-1,2-dihydro-2,2,4-trimethylquinoline)

Ethanol
$C = 0.021$ g/liter
$d = 5.01$ mm

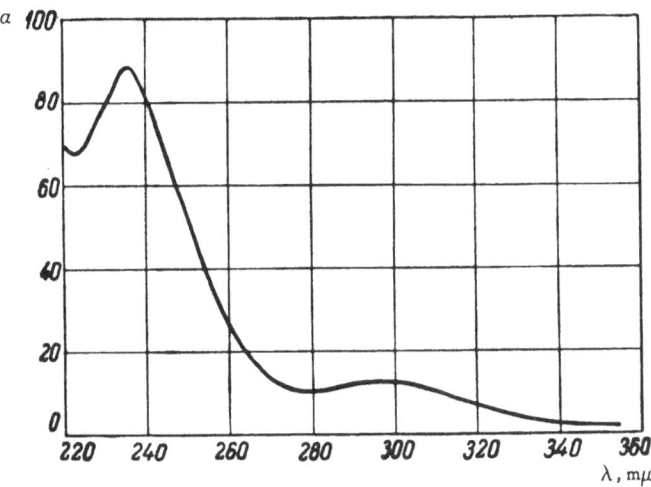

Fig. 102. Agerite Resin D

Ethanol
$C = 0.91$ g/liter
$d = 0.214$ mm

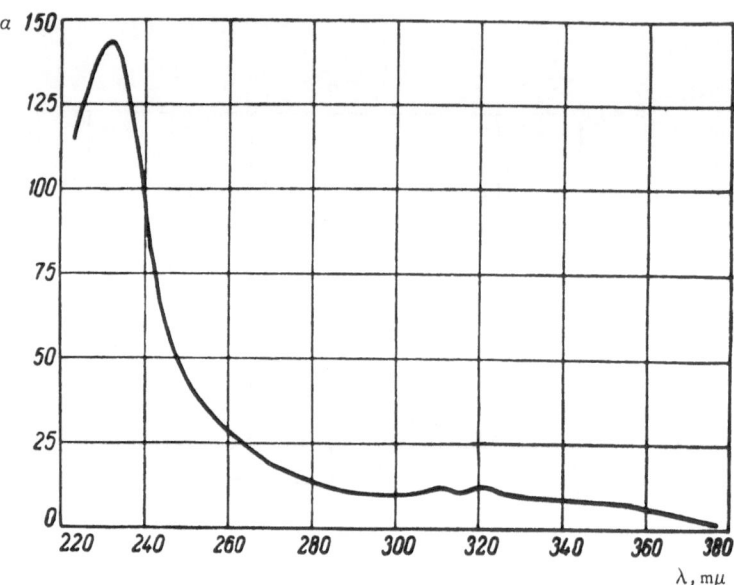

Fig. 103. Flectol B

Ethanol
C = 0.1464 g/liter
d = 0.509 mm

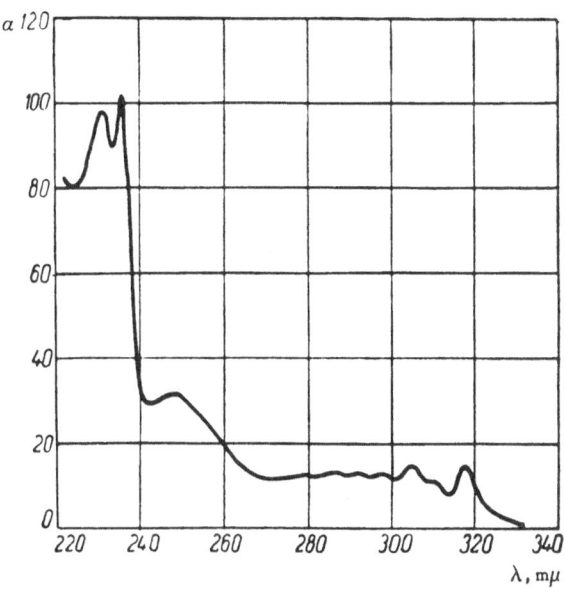

Fig. 104. Antox

Ethanol
$C = 0.152$ g/liter
$d = 1.016$ mm

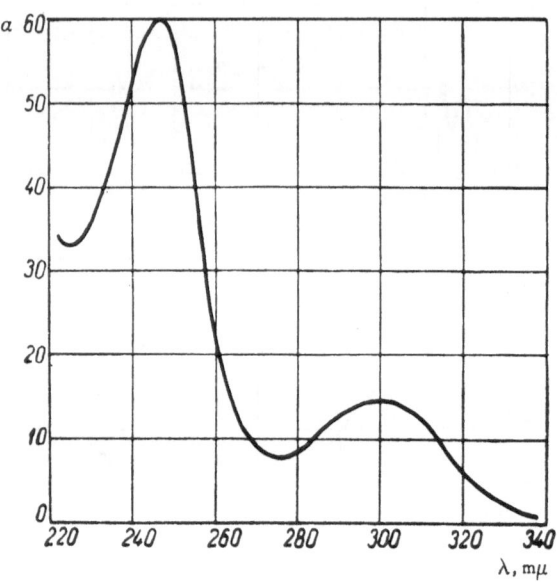

Fig. 105. Antioxidant I-18

Ethanol
$C = 0.922$ g/liter
$d = 0.117$ mm

TABLE 12

SPECTRAL CHARACTERISTICS OF VARIOUS ANTIOXIDANTS

Figure	Antioxidant	Solvent	λ_{max}, mμ	a	ϵ
106	Mercaptobenzimidazole (2-benzimidazolethiol)	Ethanol	246 305	90 194	13,400 28,900
107	Flexamine (mixture of 65% of product of condensation of a diarylamine with a ketone and 35% of N,N-diphenyl-p-phenylenediamine)	Chloroform	295	71.5	—
108	Akroflex C (mixture of 65% of Neozone A and 35% of N,N'-diphenyl-p-phenylenediamine)	Ethanol	252 304	63 50	— —
109	Santoflex BX (mixture of the antioxidant Santoflex B and N,N'-diphenyl-p-phenylenediamine)	Ethanol	234 301	37.5 45	— —

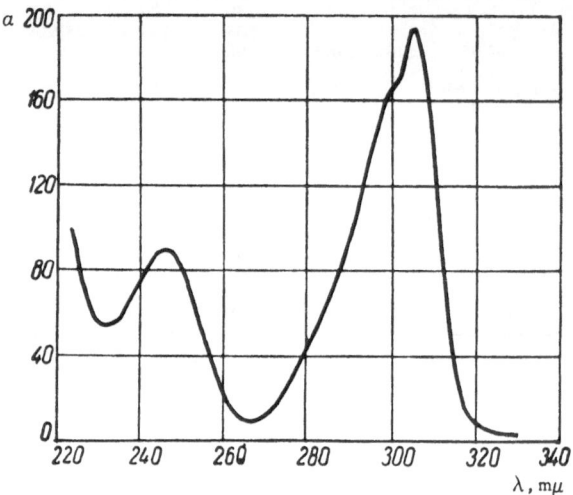

Fig. 106. Mercaptobenzimidazole (2-benzimidazolethiol)

Ethanol
$C = 0.216$ g/liter
$d = 0.214$ mm

126

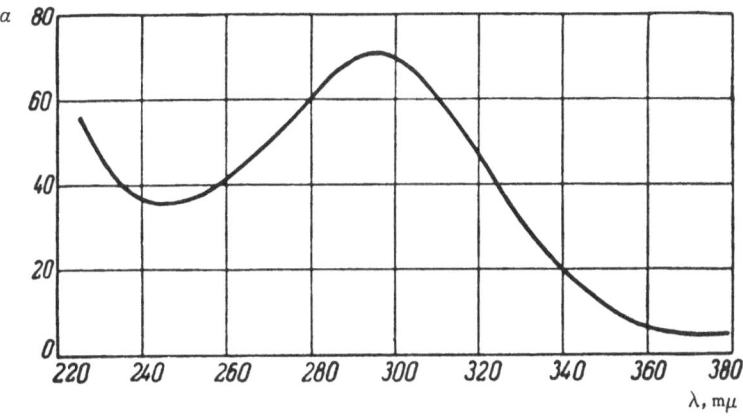

Fig. 107. Flexamine

Chloroform
$C = 0.754$ g/liter
$d = 0.117$ mm

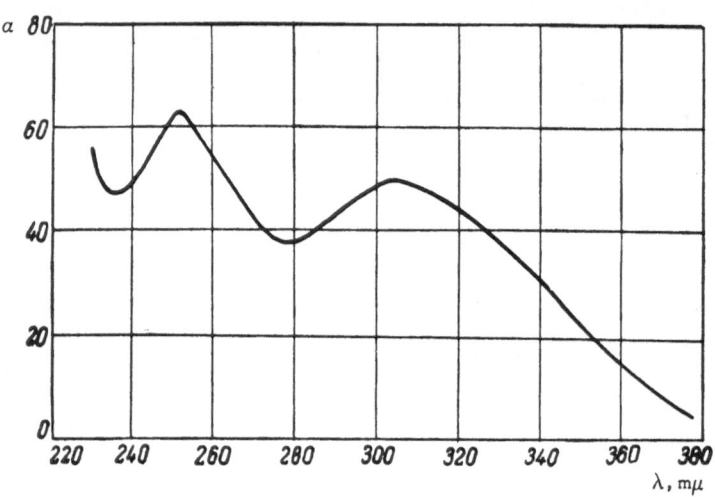

Fig. 108. Akroflex C

Ethanol
$C = 0.470$ g/liter
$d = 0.117$ mm

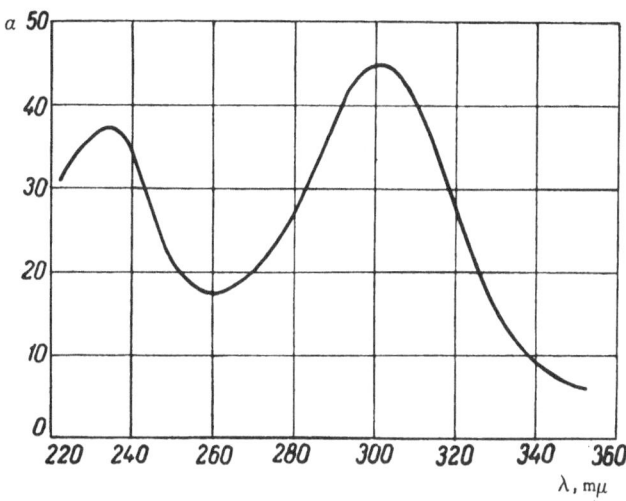

Fig. 109. Santoflex BX

Ethanol
$C = 1.50$ g/liter
$d = 0.107$ mm

V. VARIOUS SUBSTANCES USED IN SYNTHETIC RUBBER MANUFACTURE

This section contains the spectra of various substances (Table 13) used as emulsifiers and polymerization initiators, regulators, and short-stoppers in the production of synthetic rubbers by emulsion polymerization. By spectrophotometric methods of analysis both the purities of these substances and their concentrations in mixtures and solutions used in the polymerization can be determined. Moreover, some of them remain in the finished rubber, and their contents can be determined [14].

Many of the compounds listed are aromatic and their absorption spectra contain both K and B bands. The absorption spectra of a number of compounds containing a disulfide bond have particular characteristics due to the interaction of the unshared electron pair of nitrogen or oxygen with the $C=S$ bond. The disulfide bond also appears to have an effect on the latter. In alkaline alcoholic solutions the spectra of these compounds change, which makes it possible to determine them in presence of other radiation-absorbing substances by the procedure used for phenolic antioxidants.

The analysis of modified rosin used as an emulsifier is conducted with the aid of the spectra of rosin acids given here [15].

TABLE 13

SPECTRAL CHARACTERISTICS OF VARIOUS SUBSTANCES USED IN
SYNTHETIC RUBBER MANUFACTURE

Figure	Product	Solvent	λ_{max}, $m\mu$	a	ϵ
110	Nekal BXG (sodium 6,7-dibutyl-2-napthalene-sulfonate)	Ethanol	232	210	72,000
			289	20	6,900
		Water	229.5	164	56,000
			290	21	7,200
111	Leucanol(disodium 6,6′-methylenedi-2-napthalenesulfonate)	Ethanol	227	235	111,000
			275	17	7,900
112	Azolyat A (sodium alkylbenzenesulfonates)	Ethanol	262	2.30	710
			269	2.70	835
			277	2.30	710
		Water	263	2.35	725
			270	2.80	865
			277.5	2.20	680
113	Abietic acid	Ethanol	234	72	19,450
			241	77	20,800
114	Neoabietic acid	Ethanol	250	80	21,600
115	Dehydroabietic acid	Ethanol	268	2.2*	590
			275.5	2.3	620
116	Emulsifier OP-4	Ethanol	206	28.6	12,500
			225	22.9	10,000
			276	3.5	1,500
			282	3.0	1,300
117	Emulsifier OP-10	Ethanol	206	136	9,550
			225	12.5	8,750
			276	2.1	1,475
			282	1.8	1,250
118	Dipropoxide [bis-(isopropoxythiocar-bonyl)disulfide]	Ethanol	238	94	25,400
			286	46	12,400
		0.1 N ethanolic KOH	228.5	97	26,200
			304	121.5	32,800

Figure	Product	Solvent	λ_{max}, mμ	a	ϵ
119	BEK [bis(ethoxythio-carbonyl) disulfide]	Ethanol	237	92	22,250
			284	43	10,400
		0.1.N ethanolic KOH	226	89	21,550
			302	122.5	29,650
120	Thiuram D [bis-(dimethylthiocarbam-oyl) disulfide]	Ethanol	272–280	43.5	10,450
		0.1 N ethanolic KOH	255	81	19,500
			285	75	18,000
121	Thiuram E [bis-(diethylthiocarbam-oyl) disulfide]	Ethanol	216.5	73	21,600
		0.1 N ethanolic KOH	259	71.5	21,150
			287	64	18,950
122	Sodium diethyldithio-carbamate	Ethanol	253.5	79	13,500
			287	85	14,500
123	Hydroquinone	Ethanol	225	55	6,050
			294	26	2,850**
		0.1 N ethanolic KOH	209	75	8,250
			242	62	6,800
			320	24	2,650
124	Cumene hydroper-oxide (a,a-dimethyl-benzyl hydroperoxide)	Ethanol	246	1.90	290
			251	2.00	305
			257	2.05	310
			264	1.45	220

*In alcohol λ_{max}= 268 mμ and 275.5 mμ, a = 2.3 and 2.2 [16].

In alcohol λ_{max}= 275 mμ , a = 2.4 [17].

**In alcohol λ_{max}= 294 mμ, ϵ = 3,100 [18].

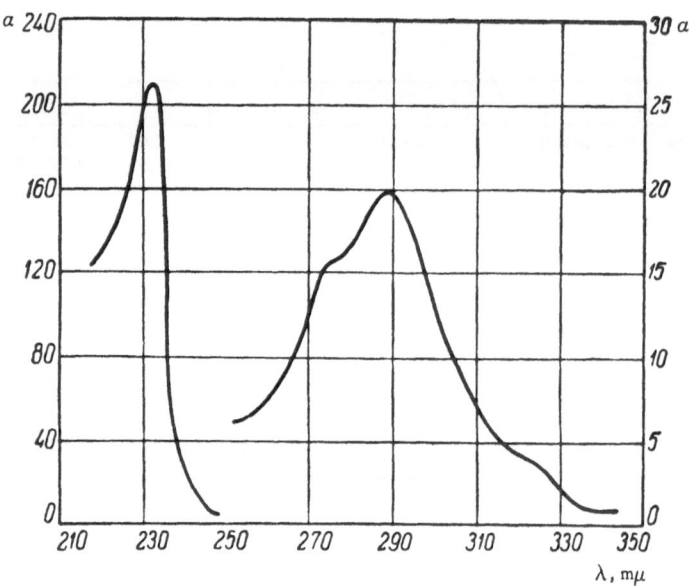

Fig. 110. Nekal BXG (sodium 6,7-dibutyl-2-naphthalenesulfonate)

C₄H₉ — [naphthalene ring] — SO₃Na

Ethanol
C = 1.66 g/liter
d = 0.049 and 0.509 mm

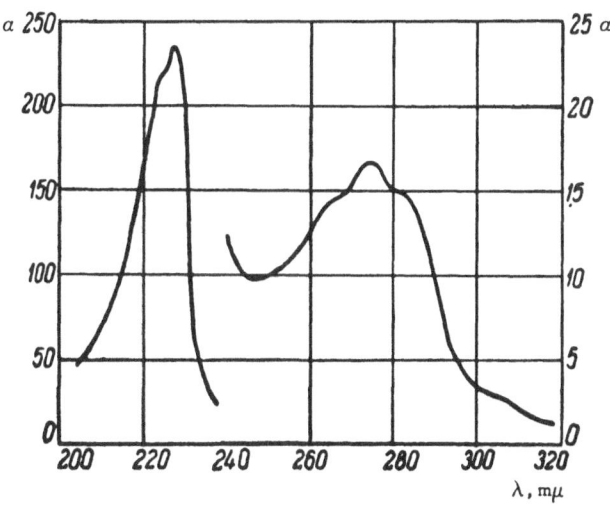

Fig. 111. Leucanol (disodium 6,6′-methylenedi-2-naphthalenesulfonate)

Ethanol
C = 0.56 g/liter
d = 0.058 and 0.506 mm

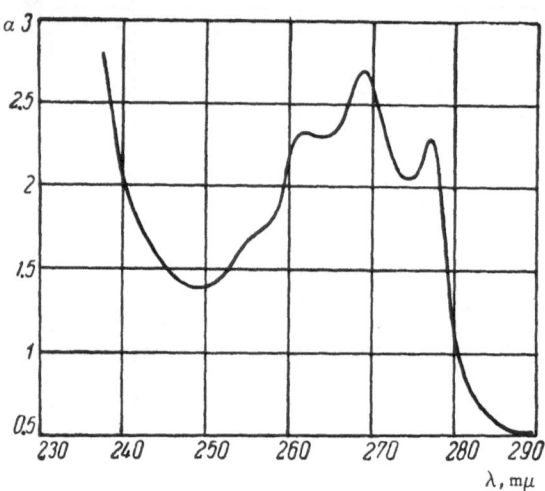

Fig. 112. Azolyat A (sodium alkylbenzenesulfonates)

C₅H₁₁—⟨benzene⟩
C₅H₁₁—⟨benzene⟩—SO₃Na   +   C₁₀H₂₁—⟨benzene⟩—3O₃Na   Ethanol
                                                             C = 0.20 g/liter
      ~ 60%   ~ 40%   d = 9.99 mm

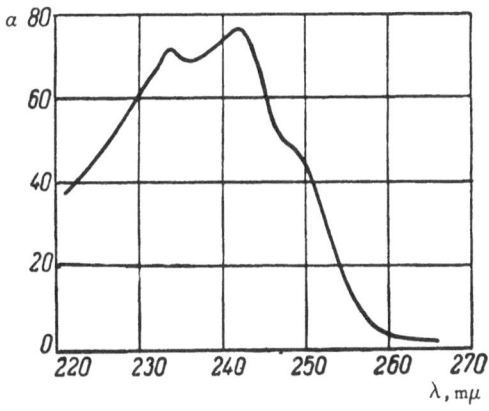

Fig. 113. Abietic acid

Ethanol [24]

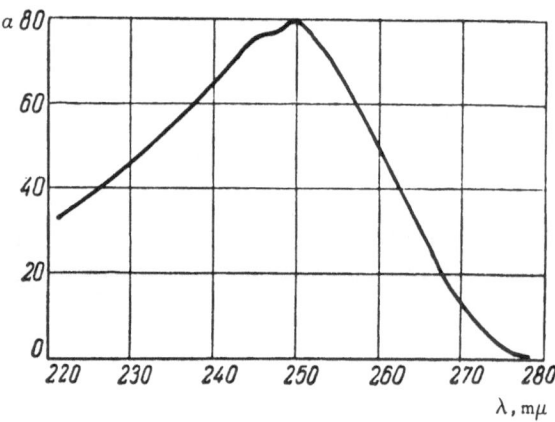

Fig. 114. Neoabietic acid

CH₃ COOH

Ethanol [24]

138

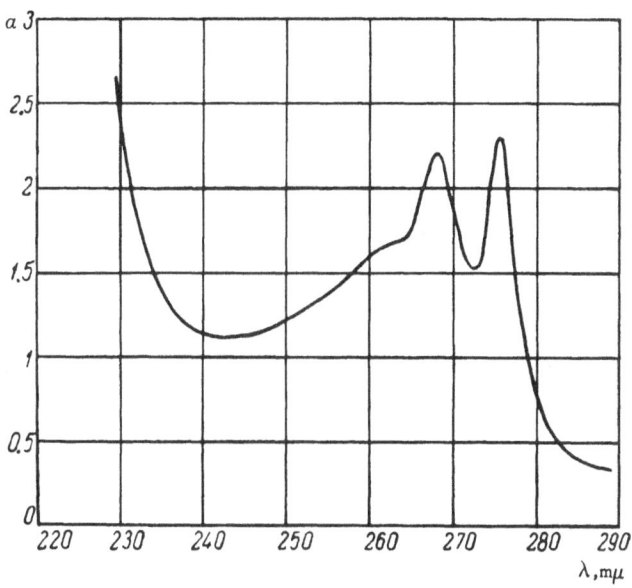

Fig. 115. Dehydroabietic acid

Ethanol
$C = 1.034$ g/liter
$d = 1.012$ mm

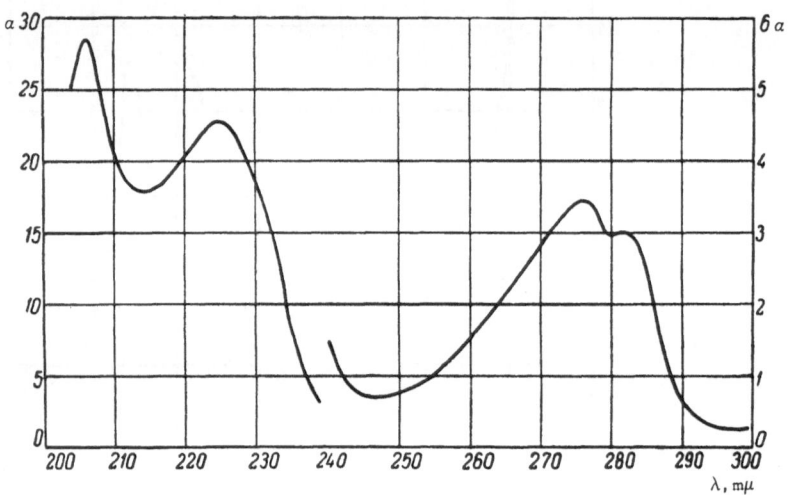

Fig. 116. Emulsifier OP-4

RC$_6$H$_4$(OCH$_2$CH$_2$)$_4$OH

Ethanol
$C = 10.2$ g/liter
$d = 0.112$ and 0.210 mm

140

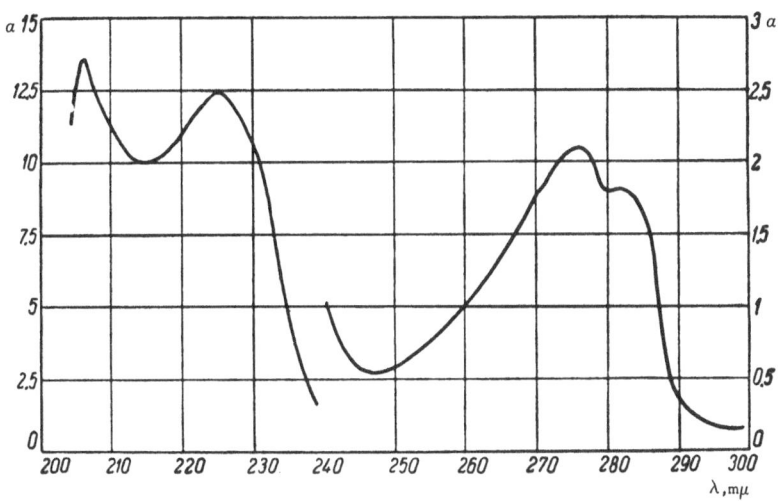

Fig. 117. Emulsifier OP-10

$RC_6H_4(OCH_2CH_2)_{10}OH$

Ethanol
$C = 10.0$ g/liter
$d = 0.112$ and 0.210 mm

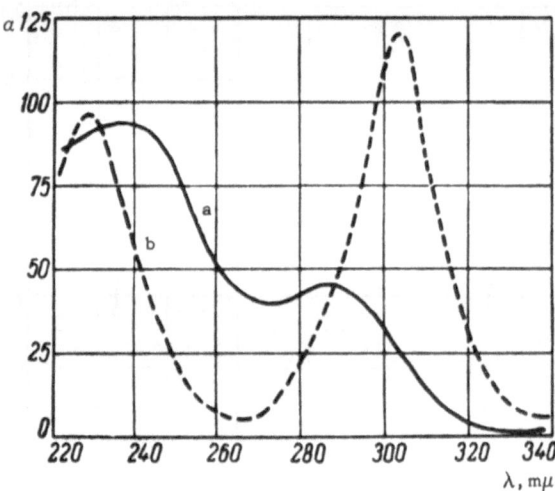

Fig. 118. Dipropoxide [bis(isopropoxythiocarbonyl) disulfide]

$(CH_3)_2CHOCSSCOCH(CH_3)_2$
|| ||
S   S

Ethanol (a), 0.1 N ethanolic KOH (b)
$C = 1.50$ g/liter
$d = 0.049$ mm

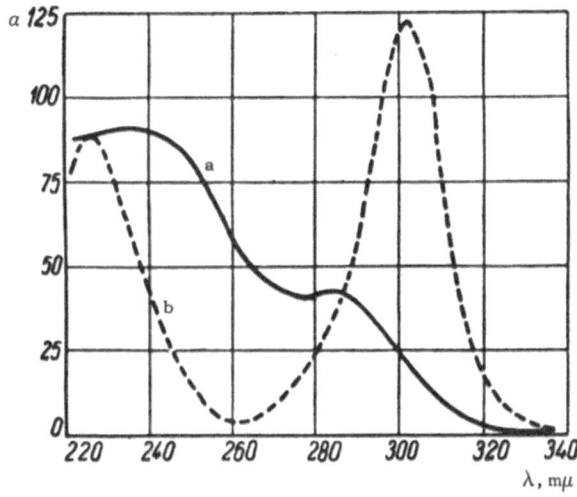

Fig. 119. BEK [bis(ethoxythiocarbonyl) disulfide]

$C_2H_5\,OCSSCOC_2H_5$
   ‖   ‖
   S   S

Ethanol (a), 0.1 N ethanolic KOH (b)
$C = 1.58$ g/liter
$d = 0.049$ mm

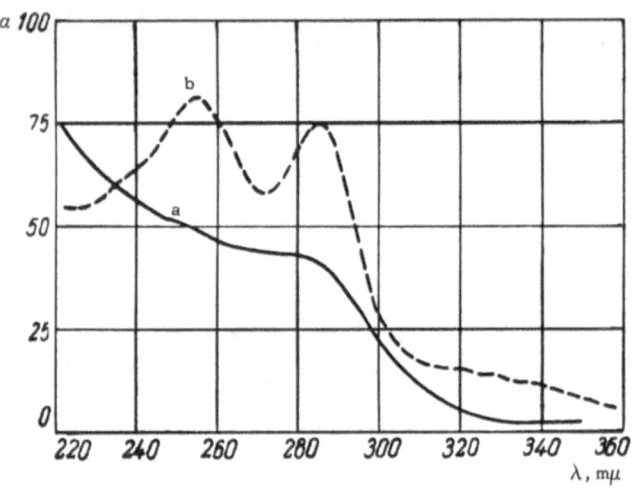

Fig. 120. Thiuram D [bis (dimethylthiocarbamoyl) disulfide]

(CH₃)₂NCSSCN (CH₃)₂
$\quad\quad$ ‖ $\quad$ ‖
$\quad\quad$ S $\quad$ S

Ethanol (a), 0,1 N ethanolic KOH(b)
C = 0.98 g/liter
d = 0.107 mm

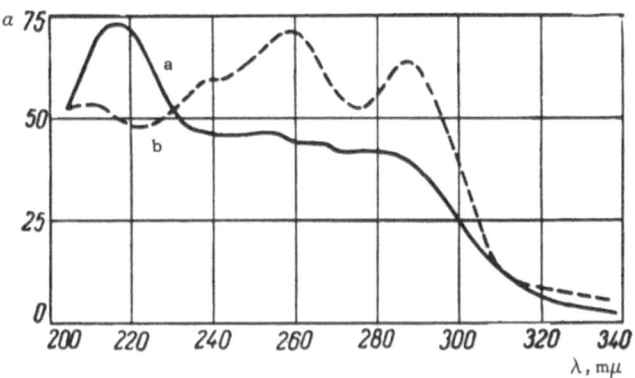

Fig. 121. Thiuram E [bis(diethylthiocarbamoyl) disulfide]

$(C_2H_5)_2NCSSCN (C_2H_5)_2$
$\quad \quad \overset{\|}{S} \quad \overset{\|}{S}$

Ethanol (a), 0.1 N ethanolic KOH (b)
$C = 0.97$ g/liter
$d = 0.112$ mm

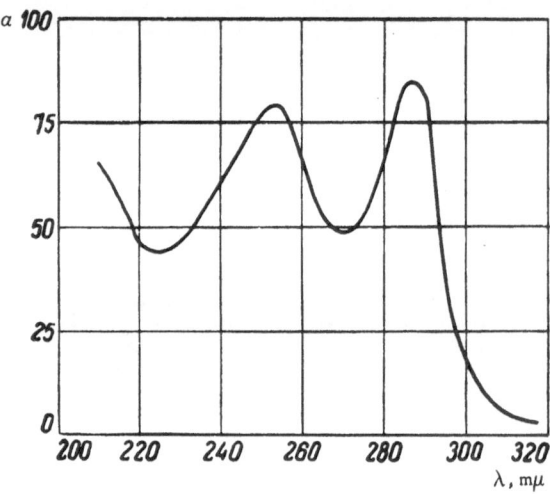

Fig. 122. Sodium diethyldithiocarbamate

$(C_2H_5)_2NCSNa$
‖
S

Ethanol
$C = 1.07$ g/liter
$d = 0.058$ mm

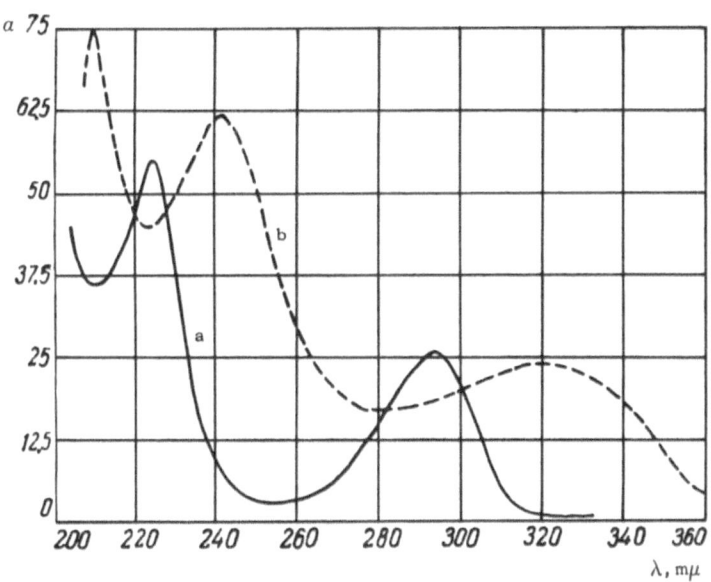

Fig. 123. Hydroquinone

Ethanol (a), 0.1 N ethanolic KOH (b)
$C = 0.884$ g/liter
$d = 0.210$ mm

147

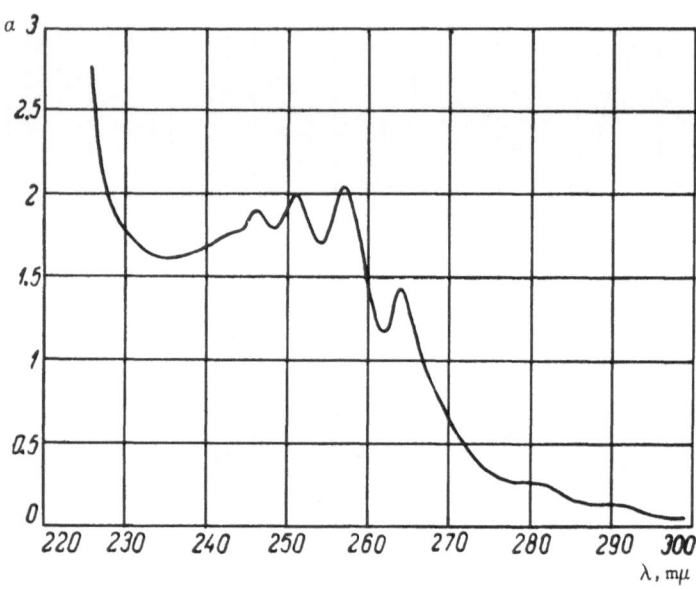

Fig. 124. Cumene hydroperoxide (a,a-dimethylbenzyl hydroperoxide)

CH₃
|
C₆H₅ – C – O – OH   Ethanol
| C = 36.6 g/liter
CH₃ d = 0.049 mm

VI. VARIOUS SUBSTANCES MET AS INTERMEDIATE PRODUCTS, BY-PRODUCTS, AND IMPURITIES

The acetylene derivatives whose spectra are given here (Table 14) are either intermediate products used in the production of chloroprene rubber (butenyne) or impurities formed in the course of syntheses (1,5-hexadien-3-yne, ethynylbenzene, butadiyne). A spectrophotometric method has been devised for the determination of the latter as impurities [19].

All these compounds show very strong K bands due to the conjugation of a triple bond with a double bond or a benzene ring. When there are two conjugated triple bonds (butadiyne) the strength of the band is greatly reduced.

This section includes also spectra of by-products formed in the synthesis of monomers and other substances used in synthetic rubber production, and also of substances formed by the decomposition of compounds used in polymerization processes. Spectral analysis is used both to determine the purities of compounds [20], and also to determine the polymerization mechanism from the identity of the products of the decomposition of compounds taking part in this process.

TABLE 14

SPECTRAL CHARACTERISTICS OF VARIOUS SUBSTANCES MET AS
INTERMEDIATE PRODUCTS, BY-PRODUCTS, AND IMPURITIES

Figure	Compound	Solvent	λ_{max}, mμ	a	ϵ
125	Vinylacetylene (butenyne)	Methanol	218	266	13,800*
			227	205	10,650
126	Divinylacetylene (1,5-hexadien-3-yne)	Methanol	252	188	14,700
			265	172	13,400
127	Phenylacetylene (ethynylbenzene)	Methanol	234.5	256	26,100
			245	229	23,350
128	Diacetylene (butadiyne)	Methanol	223.5	9.5	475
			234	10.5	525
			246	5.0	250
129	Acetaldehyde	Water	276.5	0.17	7.5
130	Butyraldehyde	Water	282	0.15	10.8
131	Crotonaldehyde	Ethanol	219	275	17,250²*
132	Acetone	Water	264	0.3	17.4
133	Cyclopentadiene	Isopentane	239	45	3,000³*
134	Benzoic acid	Ethanol	272	7.50	915
			280	6.20	755
		0.1 N ethanolic KOH	262	4.80	585
			269	4.50	550
			276	3.25	395
135	Acetophenone	Water	245	93	11,150
136	a,a-Dimethylbenzyl alcohol	Water	250.5	1.45	200
			256	1.55	210
137	Chlorobenzene	Ethanol	251	1.20	135
			257.5	2.00	225
			264	2.55	285
			271	1.85	210
138	p-Dichlorobenzene	Ethanol	258.5	1.85	270
			265	2.75	425
			272.5	3.70	545
			280.5	3.05	470
139	o-Dichlorobenzene	Ethanol	262.5	2.05	320
			269.5	2.55	375
			277	1.95	305
140	5-Ethyl-2-picoline	Ethanol	268	50	6,000
141	2-Nitropropene	Ethanol	222	48	4,200⁴*

*λ_{max} = 219 mμ, = 6400 [7].
²* In alcohol λ_{max} = 217 mμ, ϵ = 15,650 [21].
³* In hexane λ_{max} = 238.5 mμ, ϵ = 3400 [22].
⁴* In alcohol λ_{max} = 225 mμ, ϵ = 3300 [23].

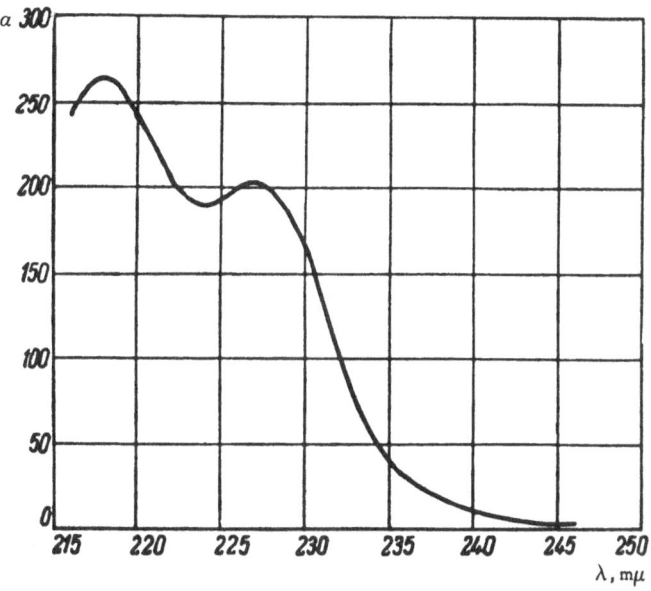

Fig. 125. Vinylacetylene (butenyne)

$$CH \equiv C - CH = CH_2$$

Methanol
$C = 1.12$ g/liter
$d = 0.049$ mm

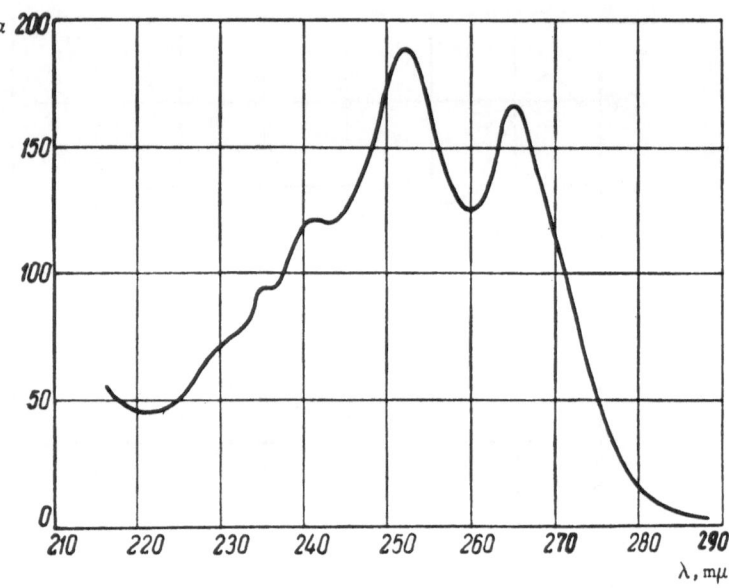

Fig. 126. Divinylacetylene (1,5-hexadien-3-yne)

$CH_2 = CH - C \equiv C - CH = CH_2$

Methanol
$C = 1.48$ g/liter
$d = 0.049$ mm

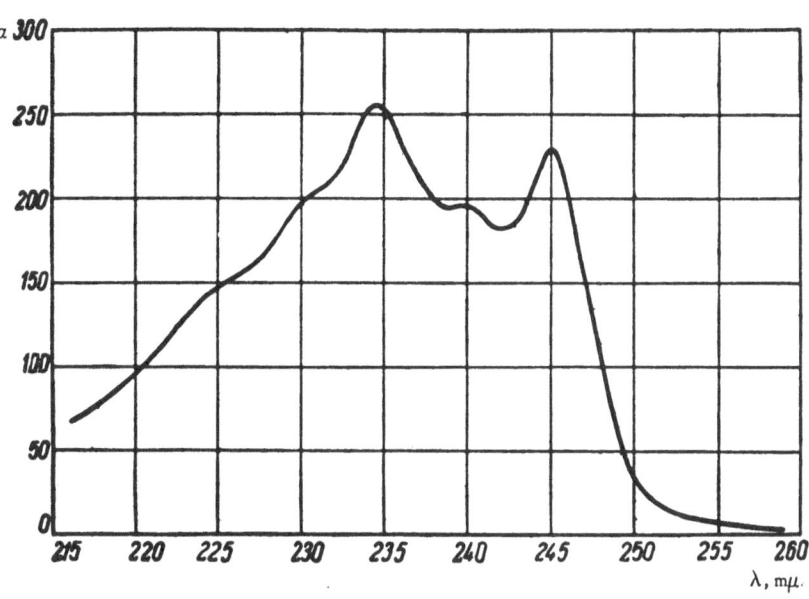

Fig. 127. Phenylacetylene (ethynylbenzene)

$CH \equiv C - C_6H_5$

Methanol
$C = 0.96$ g/liter
$d = 0.049$ mm

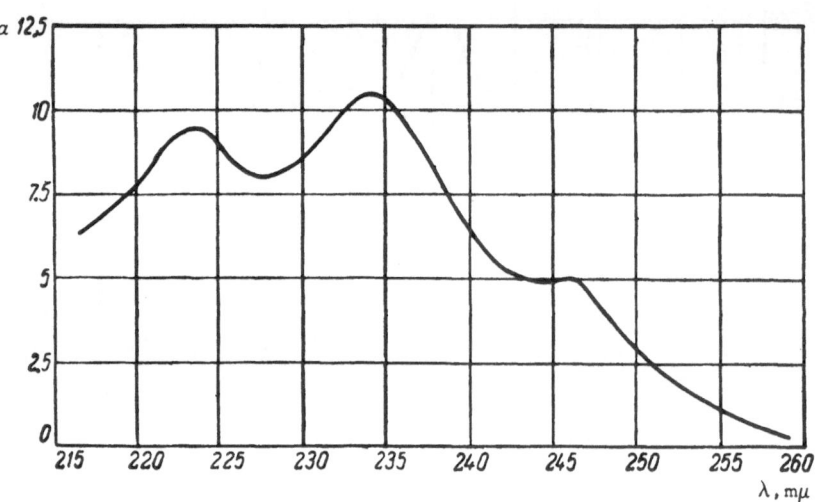

Fig. 128. Diacetylene (butadiyne)

CH≡C–C≡CH

Methanol
C = 8.16 g/liter
d = 0.049 mm

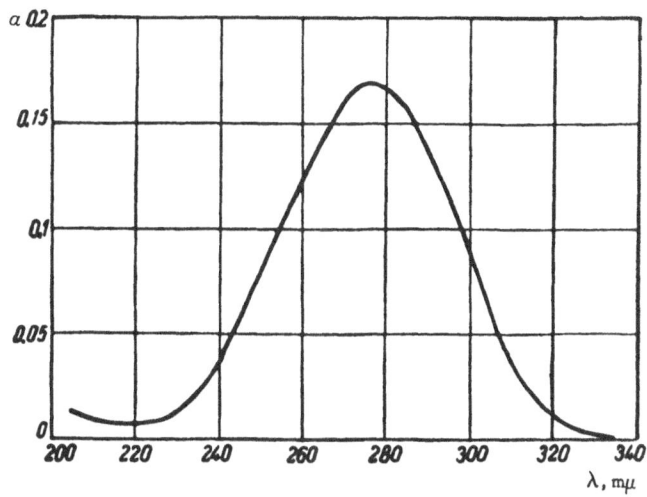

Fig. 129. Acetaldehyde

$CH_3CH = O$

Water
$C = 32.5$ g/liter
$d = 1.004$ mm

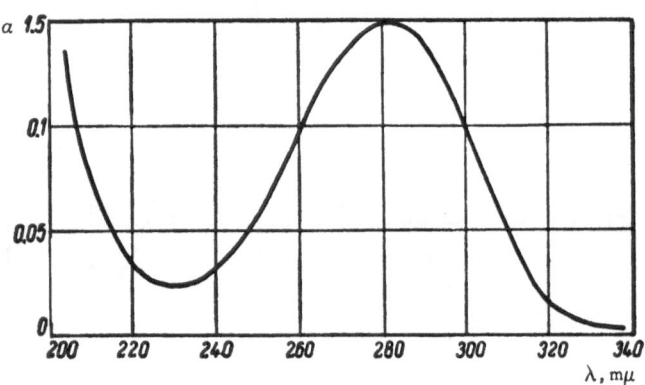

Fig. 130. Butyraldehyde

$CH_3CH_2CH_2CH=O$

Water
$C = 5.04$ g/liter
$d = 9.99$ mm

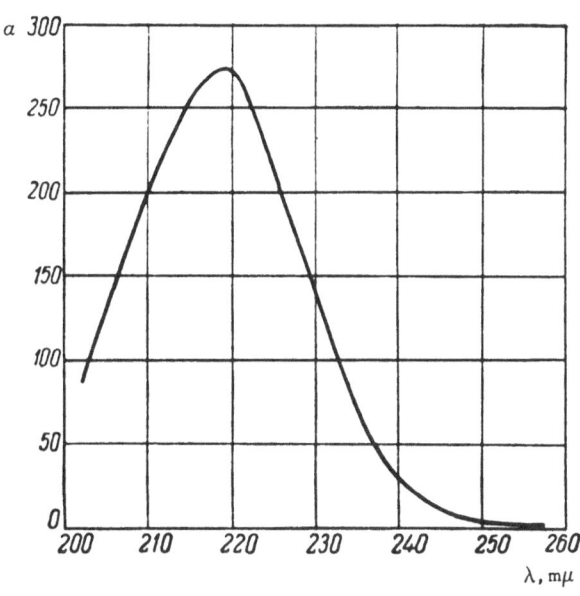

Fig. 131. Crotonaldehyde

$CH_3CH = CHCHO$

Ethanol
$C = 0.804$ g/liter
$d = 0.058$ mm

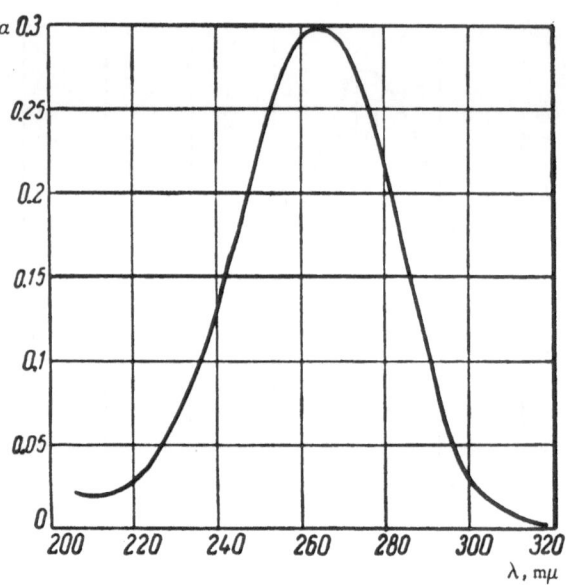

Fig. 132. Acetone

$(CH_3)_2 C = O$

Water
$C = 4.85$ g/liter
$d = 5.00$ mm

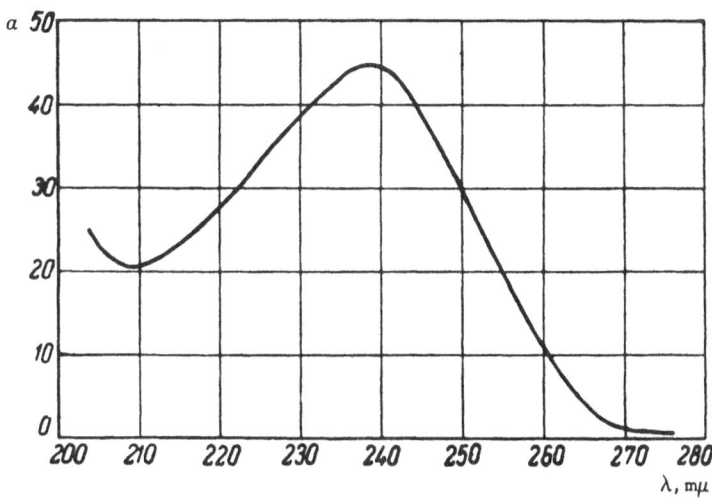

Fig. 133. Cyclopentadiene

CH – CH
‖   ‖
CH   CH   Isopentane
 \   /   C = 2.27 g/liter
  CH₂   d = 0.112 mm

159

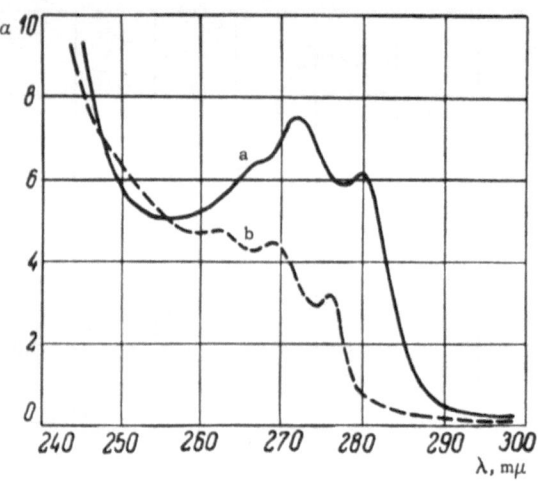

Fig. 134. Benzoic acid

C₆H₅ COOH Ethanol (a), 0.1 N ethanolic KOH (b)
$C = 10.0$ g/liter
$d = 0.107$ mm

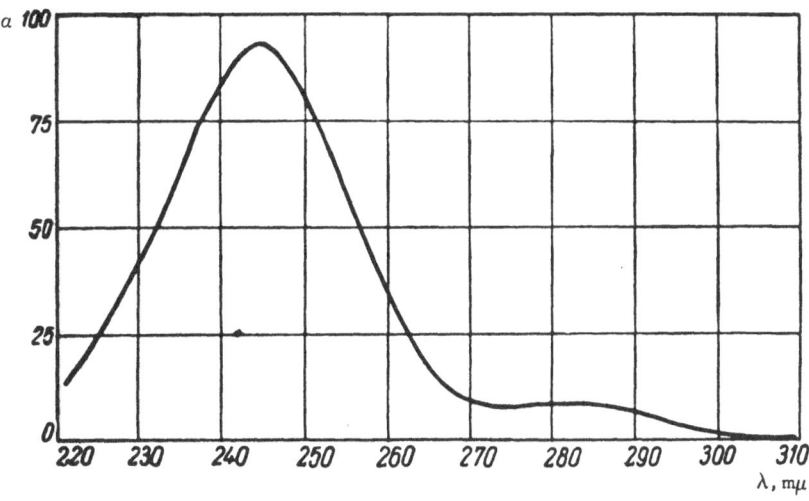

Fig. 135. Acetophenone

$$C_6H_5\,C \overset{\nearrow O}{\underset{\searrow CH_3}{}}$$

Water
$C = 0.33$ g/liter
$d = 0.212$ mm

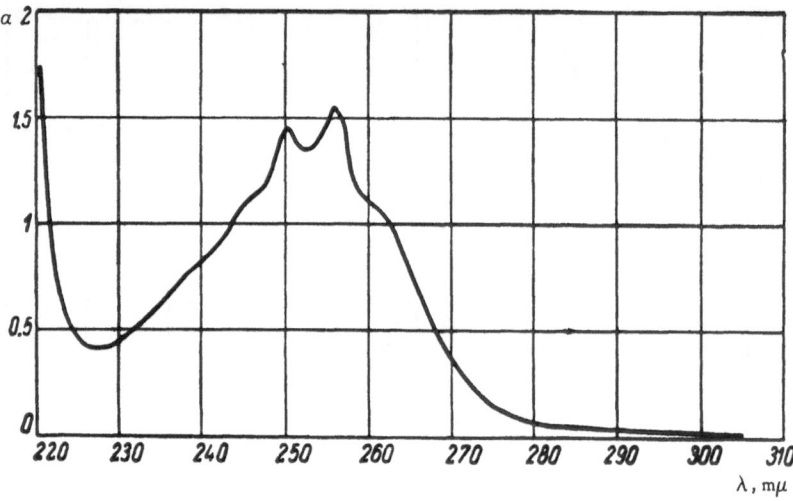

Fig. 136. α,α-Dimethylbenzyl alcohol

$$C_6H_5 - \overset{\overset{\displaystyle CH_3}{|}}{\underset{\underset{\displaystyle CH_3}{|}}{C}} - OH$$

Water
$C = 1.05$ g/liter
$d = 5.00$ mm

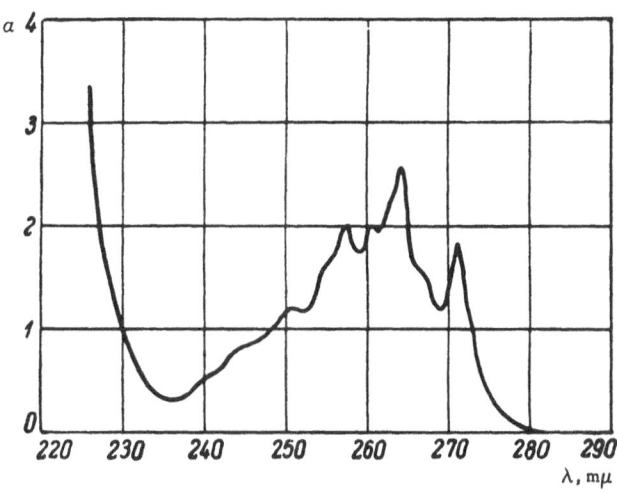

Fig. 137. Chlorobenzene

Ethanol
$C = 25.9$ g/liter
$d = 0.049$ mm

$C_6H_5 Cl$

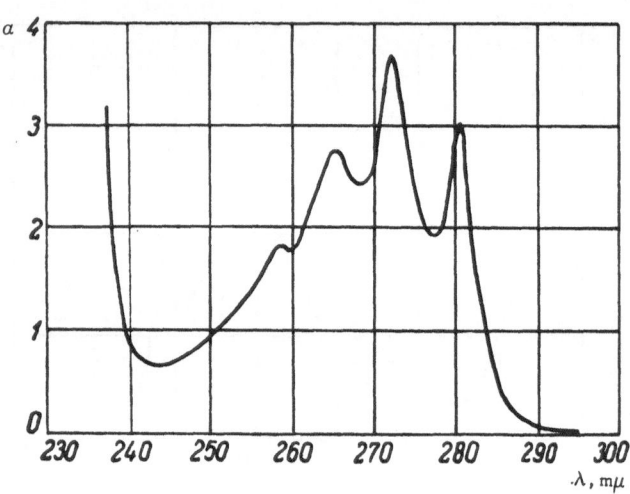

Fig. 138. *p*-Dichlorobenzene

Ethanol
$C = 38.7$ g/liter
$C_6H_4Cl_2$ $d = 0.049$ mm

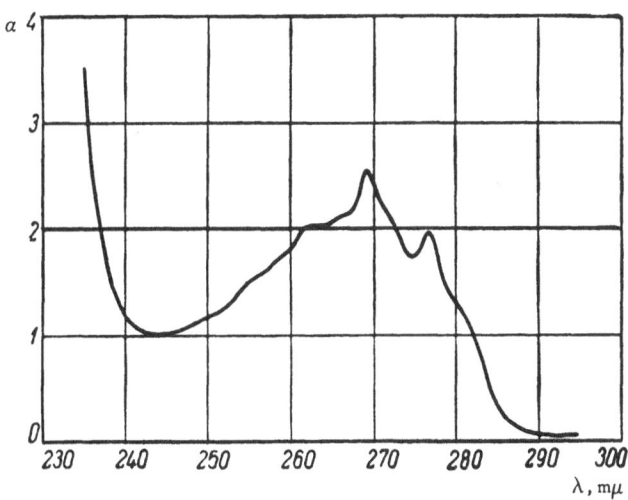

Fig. 139. o-Dichlorobenzene

$C_6H_4Cl_2$

Ethanol
C=41.8 g/liter
d=0.049 mm

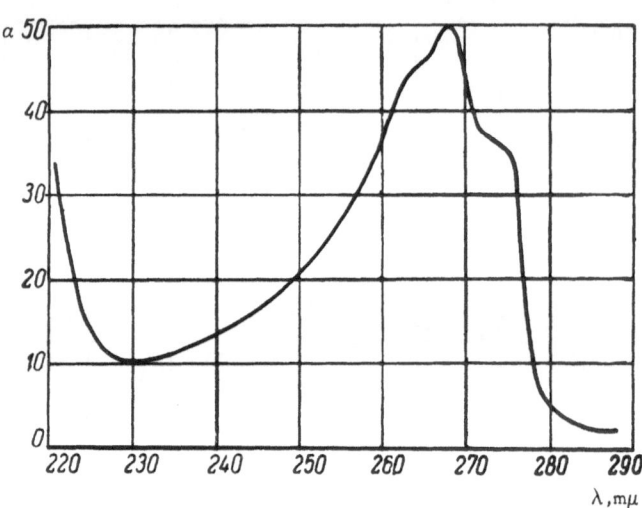

Fig. 140. 5-Ethyl-2-picoline

Ethanol
C = 1.58 g/liter
CH₃C₅H₃NC₂H₅   d = 0.049

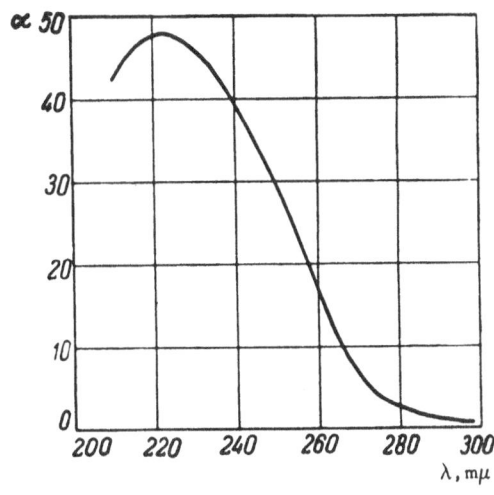

Fig. 141. 2-Nitropropene

$$CH_2 = C - CH_3$$
$$|$$
$$NO_2$$

Ethanol
$C = 2.15$ g/liter
$d = 0.112$ mm

LITERATURE CITED

1. M. M. Kusakov, N. A. Shimanko, and M. V. Shishkina, Ultraviolet Absorption Spectra of Aromatic Hydrocarbons, (Izd. AN SSSR, 1963).
2. A. Gillam and E. S. Stern, Introduction to Electronic Absorption (St. Martins), [Russian translation: Electronic Absorption Spectra of Organic Compounds (IL, 1957)]; E. A. Braude and F. C. Nachod, Determination of Organic Structures by Physical Methods, Ch. 4, N.Y., 1955.
3. V. M. Peshkova and M. I. Gromova, Practical Handbook on Spectrophotometry and Colorimetry, MGU, 1961.
4. V. N. Mironova and V. V. Zharkov, Vysokomol. soed., 2(7): 1013 (1960).
5. Smakula, Angew. Chem., 47:653 (1934).
6. A. Gillam and E. S. Stern, op. cit. [Russian translation] p. 229.
7. Ibid., p. 132.
8. Booker, Evans, and Gillam, J. Chem. Soc., 1940:1453.
9. Koch, J. Chem. Soc., 1949:387; Robertson and Matsen, J. Am. Chem. Soc., 72:5250 (1950).
10. L. M. Pyrkov, S. E. Bresler, and S. Ya. Frenkel', Zhur. Org. Khim. 29(8):2750 (1959).
11. V. S. Fikhtengol'ts and R. V. Zolotareva, Spectrophotometric Investigation and Analysis of Polyarylmethylsiloxanes, in the Collection "Physicochemical Methods for the Analysis and Investigation of Substances Used in Synthetic Rubber Manufacture," Goskhimizdat, 1961.
12. S. B. Dolgoplosk, A. L. Klebanskii, L. P. Fomina, V. S. Fikhtengol'ts, and E. Yu. Shvarts, Doklady Akad. Nank SSSR 150(4):813 (1963). [English translation: Doklady Chemistry 150:461 (1963).]
13. S. N. Borisov, A. V. Karlin, E. A. Chernyshev, and V. S. Fikhtengol'ts, Vysokomol. soed., 4(10):1507 (1962).

14. V. S. Fikhtengol'ts and R. V. Zolotareva, Spectrophotometric Methods for the Analysis of Synthetic Rubbers, in the Collection "Physicochemical Methods for the Analysis and Investigation of Substances Used in Synthetic Rubber Manufacture," Goskhimizdat, 1961.

15. V. S. Fikhtengol'ts, Zav. lab., 27(4):400 (1961). [English translation: Industrial Laboratory 27(4):402 (1961).]

16. Fieser and Campbell, J. Am. Chem. Soc., 60:159 (1938).

17. Ahlers and O'Neill, J. Oil and Col. Chem. Assoc. 37:552 (1954).

18. Morton and Stubbs, J. Chem. Soc., 1940:1349; Morton and Sawires, J. Chem. Soc., 1940:1952.

19. I. Ya. Poddubnyi, I. V. Nel'son, and R. V. Zolotareva, Spectrophotometric Method for the Determination of 1,5-Hexadien-3-yne and Butenyne Impurities, in the Collection "Physicochemical Methods for the Analysis and Investigation of Substances Used in Synthetic Rubber Manufacture," Goskhimizdat, 1961.

20. V. S. Fikhtengol'ts and R. V. Zolotareva, Spectrophotometric Control of Chlorobenzene Production, in the Collection "Physicochemical Methods for the Analysis and Investigation of Substances Used in Synthetic Rubber Manufacture," Goskhimizdat, 1961.

21. Hausser, Kuhn, Smakula, and Hoffer, Z. phys. Chem., B29:371 (1935).

22. Scheibe, Ber., 59:1333 (1926).

23. Braude, Jones, and Rose, J. Chem. Soc., 1947:1104.

24. S. Harrison and T. Sanderson, J. Am. Chem. Soc., 70:334 (1948).